HYDROPOLITICS IN THE MEKONG REGION

Water, Power, and the Dialectics of Change

Edited by

Kim Geheb and Diana Suhardiman

Leiden University Press

This book will be made open access within three years of publication thanks to Path to Open, a programme developed in partnership between JSTOR, the American Council of Learned Societies (ACLS), University of Michigan Press, and The University of North Carolina Press to bring about equitable access and impact for the entire scholarly community, including authors, researchers, libraries, and university presses around the world. Learn more at https://about.jstor.org/path-to-open/.

Environmental Governance series, volume 2

Cover design: Andre Klijsen
Cover illustration: Diana Suhardiman
Layout: Crius Group, Hulshout
Printer: Scanlaser bv, Zaandam

ISBN 9789087284206
e-ISBN 9789400604636 (e-PDF)
e-ISBN 9789400605657 (e-PUB)
https://doi.org/10.24415/9789087284206
NUR 934

Contact EU General Product Safety Regulation (GPSR): productsafety@lup.nl
Printed and bound by CPI Group (UK) Ltd, Croydon, CR0 4YY

Table of Contents

List of Figures

List of Tables

Acronyms, abbreviations, and initialisations

3S	Sesan, Srepok, and Sekong (rivers)
ADB	Asian Development Bank
BOT	Build-Own-Transfer
BRI	Belt and Road Initiative
CEIC	China Energy Investment Group
CEO	Chief Executive Officer
CFs	Community Fisheries
CITES	Convention on International Trade in Endangered Species of Wild Fauna and Flora
CNMC	Cambodian National Mekong Committee
CSE	Chaleun Sekong Energy
CSO	Civil Society Organisation
DP	Development Partner
EdL	Électricité du Lao
EdL-GEN	EdL-Generation (Laos)
EGAT	Electricity Generating Authority of Thailand
EIA	Environmental Impact Assessment
EVN	Electricity Vietnam
FCA	Fisheries Conservation Areas
GDP	Gross Domestic Product
GED	Gulf Energy Development PLC
GMS	Greater Mekong Subregion
GOL	Government of Laos
GW	Gigawatts
GWh	Gigawatt Hours
ha	Hectares
ICCA	Indigenous and Community Conserved Areas
IMC	Interim Mekong Committee
IUCN	International Union for the Conservation of Nature
IWRM	Integrated Water Resources Management
JC	Joint Council (of the MRC)
KESAN	Karen Environmental and Social Action Network
LMB	Lower Mekong Basin
LMC	Lancang–Mekong Cooperation
LPRP	Lao People's Revolutionary Party

LSS2	Lower Sesan 2 (hydropower plant in Cambodia)
MAFF	Ministry of Agriculture, Forestry and Fisheries (Cambodia)
MC	Mekong Committee
MDM	Mekong Dam Monitor
MOE	Ministry of the Environment (Cambodia)
MONRE	Ministry of Natural Resources and Environment (Laos and Vietnam)
MOWRAM	Ministry of Water Resources and Meteorology (Cambodia)
MPI	Ministry of Planning and Investment (Laos)
MW	Megawatts
MWh	Megawatt Hours
MRC	Mekong River Commission
NGO	Non-Governmental Organisation
NMC	National Mekong Committee
NT2	Nam Theun 2 (hydropower plant in Laos)
PDP	Power Development Plan
PNPCA	Procedures for Notification, Prior Consultation and Agreement
SOE	State-Owned Enterprise
TAR	Tibet Autonomous Region
THPC	Theun-Hinboun Power Company
TSA	Tonle Sap Authority
TSBR	Tonle Sap Biosphere Reserve
TSL	Tonle Sap Lake
UNECAFE	United Nations Economic Commission for Asia and the Far East
UNESCO	United Nations Educational, Scientific and Cultural Organization
VMD	Vietnamese Mekong Delta
WWF	World Wildlife Fund
YPG	Yunnan Provincial Government

CHAPTER 1

The bliss of development: Mekong hydropolitics and the power of narrative

Kim Geheb[1] and Diana Suhardiman[2]

Abstract

This chapter introduces the volume's central argument: that the Mekong has been reconstituted through hydropolitical processes that pattern space, determining what flows and what is blocked, who benefits, and who bears costs. Drawing on political ecology, hydrosocial theory, and critical work on development narratives, we demonstrate that this reconfiguration is the deliberate outcome of power relations operating across multiple scales. Narratives do not merely describe these transformations; they constitute the hydropolitical spaces within which they unfold. Understanding Mekong hydropolitics requires attention to three dimensions: how power patterns hydrosocial systems, how narratives produce the spaces within which development proceeds, and how costs and benefits are inequitably distributed. We conclude that recognising the political nature of these ostensibly technical decisions is the essential first step toward more just hydropolitical futures.

Keywords: Mekong, hydropolitics, political ecology, hydropower, narrative

Introduction: The Mekong hydropolitical

The Mekong River presents a striking paradox. It remains one of the world's most biologically productive river systems, supporting the largest inland fishery on Earth and providing food security and livelihoods for over 60 million people across six nations. Yet it is simultaneously an intensely contested waterscape, subject to competing claims, accelerating infrastructure development, and deep ecological crises. Since the early 1990s, the basin has been transformed from a largely free-flowing system to one increasingly impounded by large hydropower dams, with profound consequences for river hydrology, ecology, riparian livelihoods, and water governance.

[1] Landscape Alliance, Nairobi, Kenya.

[2] Royal Netherlands Institute of Southeast Asian and Caribbean Studies and Leiden University, Leiden, The Netherlands.

This volume intervenes in debates about the Mekong not as a survey of regional issues but as an analytical contribution to understanding how its hydropolitics operate. By 'hydropolitics' we mean the nexus of water, power, and politics that produces spatial configurations of access, exclusion, and accumulation. This framing draws attention to the ways in which rivers are not merely natural systems to be managed but are actively produced through social, political, and economic processes that determine who benefits from water and its associated resources, and who bears the costs of its transformation.

The chapters assembled here, spanning empirical cases from China's Yunnan Province to Vietnam's Mekong Delta, from the Salween in Myanmar to the Tonle Sap in Cambodia, collectively demonstrate a central argument: the Mekong is perpetually being reconstituted, a process driven by different elites and their associated spheres of influence, at different times and at different places. These are the constantly shifting outcomes of hydropolitical processes operating across multiple scales, from village fishing communities to provincial governments, from national energy planners to transnational investment networks—and, at the centre of it all—usually relegated in the competitive regional narrative space, local communities, and natural value (Manorom and Geheb, this volume).

Our argument proceeds through three interrelated dimensions. First, we examine how power patterns the Mekong as a hydropolitical space—producing particular spatial configurations. Second, we analyse the role of dominant narratives in constituting the spaces within which hydropolitics unfolds, drawing on Roe's concept of narrative spaces. Third, we trace how these narrative spaces enable inequitable accumulation while externalising costs to peripheries, minorities, and future generations.

Constructing 'the Mekong': Region, scale, and the politics of definition

Before examining how the Mekong is being transformed, we must ask what 'the Mekong' is. The Mekong is not a self-evident geographical object but a produced region, constituted through colonial cartography, Cold War geopolitics, development imaginaries, and contemporary infrastructure networks. Drawing on Massey (1994, 2012), we understand space as inherently fluid: emerging through interrelations across scales, constituting a sphere of multiplicity where distinct trajectories mingle, and constantly under construction through material practices and relations. How space is organised is inherently political, shaping which interactions and futures become possible.

Most contemporary constructs gather around a river → basin → region scaling. But this conceals the politics of scalar construction itself. The basin is not a discovery but a production. When the Mekong Committee (and later, the Mekong River Commission (MRC)) adopted the basin as their unit of analysis, this was a political act that enrolled hydrology to legitimate a particular scalar framing (Wong, 2010). Bakker (1999) makes precisely this point: the Asian Development Bank and MRC 'discursively formed' the Mekong basin as a 'natural unit' and 'underutilised resource'.

Sneddon and Fox (2006) identified three dimensions through which the Mekong basin has been defined as a region: geopolitical, economic, and biophysical. Each produces a different 'Mekong' with different implications for governance. These Mekongs do not simply coexist; they compete for authority to define what counts as legitimate knowledge and appropriate governance. Middleton and Allouche (2016), drawing on Magee (2006), argue that the Mekong has been reconstituted from a watershed to a powershed—a space defined by the flow of electrical power along transmission lines rather than by the flow of water downhill.

We propose to understand the Mekong not as a single powershed with definable boundaries but as plural terrains on which multiple, overlapping, partially compatible, and partially contradictory projects of domination are pursued. The boundaries of the Mekong Region are fuzzy—not as a bug, but as a feature. They allow actors to shift between frames, to claim sovereignty when convenient and cooperation when useful, to invoke science against local knowledge and local knowledge against inconvenient science, thus bringing to light how the Mekong is continuously reconstructed through the politics of knowledge production.

Key dimensions of the Mekong political-discursive object include: state territorialisation—the extension of authority into peripheries, the 'dissolution' of alternative governance forms and their 'conservation' as folklore; infrastructure as a claim—dams as material assertions of modernity; externalisation of costs—benefits concentrating among elites while costs disperse to the peripheries and future generations; procedural legitimation—institutions like the MRC that provide the appearance of governance while being stripped of regulatory power; and scalar politics—the strategic deployment of scale to pursue particular ends.

Theorising the hydropolitical: From hydrological cycle to hydrosocial relations

The theoretical architecture of this volume draws on political ecology, which foregrounds power relations, scale, and the co-production of nature and society (Robbins, 2012; Forsyth, 2003). As Harvey (1993: 25) articulates, "all ecological projects (and arguments) are simultaneously political-economic projects (and arguments) and vice versa". Political ecology insists that what appears as 'natural' is often the sedimented outcome of social processes.

Central to our approach is the hydrosocial cycle (Linton and Budds, 2014; Swyngedouw, 2004), which insists that water and society mutually constitute each other over space and time. Water is produced through particular socio-natural processes: the dam that stores monsoon flows for dry-season release, the irrigation canal that channels water to some fields and not others, and the transmission line that converts falling water into electricity consumed hundreds of kilometres away.

Political economy remains foundational. The accumulation imperative, the logic of rent-seeking, and the distribution of costs and benefits cannot be reduced to discourse. Dams are built because they generate profits for connected elites; sediment is trapped because nobody with power has an interest in its passage. We avoid the term 'developmentalist state'—in much of the Mekong Region, 'developmentalism' is a legitimising narrative rather than an achieved institutional form. Capital has proven adept at navigating ecological limits through successive adaptations: carbon trading, 'nature-based solutions', and 'sustainable hydropower'—fixes through which accumulation proceeds despite the management of ecological tension (Harvey, 2014). These also serve to greenwash, a key narrative instrument.

Power as patterning: producing hydropolitical configurations

If political ecology foregrounds power, we must be specific about what power does. We suggest that power patterns produce particular spatial configurations that may appear natural but are in fact the outcome of political decisions. This patterning operates across multiple registers: material (who gets electricity, whose land is flooded), institutional (whose rules govern), and discursive (whose accounts prevail).

The concept of patterning draws on Massey's (1993) 'power geometries'. Power does not merely advantage some actors over others; it produces the very terrain on which hydropolitics unfolds. If power only distributed benefits and burdens, redistributive politics might suffice. But if power patterns the field itself—determining what counts as a 'benefit', who qualifies as 'affected', and what constitutes 'development'—then contestation must engage at a deeper level. We use 'pattern' rather than 'demarcate' deliberately: demarcation implies boundary drawing, the separation of inside from outside. Patterning is more capacious—it includes demarcation but also gradation, arrangement, and emergence. Power does not merely draw lines; it shapes the entire configuration of flows and blockages, concentrations, and dispersals.

Consider how power patterns the Mekong's flows. Baird (this volume) examines how governments and dam operators make choices about water storage and release—choices "often presented as technical, but actually quite political". The Lancang cascade gives China the capacity to fundamentally alter downstream hydrology, but this capacity is exercised through what appears to be technical management: reservoir operation rules, power purchase agreements, and demand forecasting. Power patterns hydrology through the language of engineering (Tran and Tuan, this volume).

Power also patterns institutional space. Suhardiman et al. (2012) identify the 'scalar disconnect' in Mekong governance: regional institutions produce basin-scale analysis that cannot be translated into national policy. The Lancang–Mekong Cooperation mechanism operates at a regional scale defined by China's strategic interests, not by hydrological connectivity—part of its own 'spatial fix' (Harvey, 1981; Dwyer, 2020).

Zhang et al. (this volume) complicate narratives of Chinese state hegemony by examining subnational actors. Yunnan Province has pursued its own interests, sometimes in tension with central government policy. Scale can also be deployed from below: Bright and colleagues (this volume) show how the Karen Environmental and Social Action Network pursues a strategy of 'messing with scales'—leapfrogging the Myanmar Union Government to seek international recognition for the Salween Peace Park.

Narratives and the production of hydropolitical space

If power patterns space, it does so substantially through narrative. The field of narrative analysis is broad (Riedke and Rottenburg, 2016), but what unifies it is recognition that narratives constitute a fundamental mode of human sensemaking. As Riedke and Rottenburg (2016: 7) observe, narratives "mediate our possible ways of being in the world", penetrating into the micro-spaces of human psychology and interrelationships.

Two analytical traditions prove particularly valuable. Hajer's (1995) work demonstrates how narratives function as mechanisms through which actors are positioned and attributions of blame, responsibility, and urgency are established. For Hajer (1995: 56), narratives achieve traction through 'discursive affinity'—the ways in which "separate elements might have similar cognitive or discursive structure which suggests that they belong together". In the Mekong context, 'hydropower', 'clean energy', and 'sustainable development' achieve discursive affinity: they feel as though they belong together, even when empirical evidence suggests otherwise.

Roe's (1991, 1994, 1999) work illuminates how dominant narratives function in policy contexts characterised by uncertainty. Policy narratives serve to underwrite and stabilise "the assumptions needed for decision-making in the face of what is genuinely uncertain and complex" (Roe, 1994: 51–52). Crucially, narratives "describe scenarios not so much by telling what should happen as about what will happen" (Roe, 1994: 36–37). This is consequential: narratives are predictive, and their predictions constitute the future that they describe.

Roe's concept of *narrative space* is particularly valuable. A narrative space is the arena within which competing accounts circulate, collide, and occasionally achieve closure. We suggest that the Mekong constitutes a narrative space in which dominant narratives have achieved such hegemonic closure that alternatives struggle for traction. The 'Battery of Southeast Asia', 'the bliss of development', and 'win-win cooperation'—these are constitutive of the hydropolitical space within which decisions are made. Alternative narratives of fisheries collapse or ecological crises are positioned as marginal, as are the complaints of those who refuse modernity.

Central to narrative dominance is Molle's (2008) 'nirvana concept'—"an ideal image of what the world should tend to… a 'horizon' that individuals and societies should strive to reach". What makes nirvana concepts analytically significant is their temporal orientation. They are fundamentally *future statements* that cannot be falsified in the present. When the Lao government promises that hydropower will deliver prosperity, this claim cannot be definitively refuted until some unspecified future arrives—and when that future disappoints, the nirvana concept can always be deferred further. The horizon recedes as one approaches it.

Hegemony and the structuring of space

This brings us to Gramsci's (1971) concept of hegemony: the production of consent, the making of particular arrangements appear natural, inevitable, even desirable. For the Mekong, hegemony operates through the mundane: the assumption that 'development' requires large infrastructure; that electricity consumption is an index of progress; that states have the right to dam rivers within their borders regardless of the downstream consequences. Infrastructural development has been cast not as one option but as the only pathway.

What distinguishes hegemonic narratives from mere propaganda is their penetration into subjectivity—the ways in which they shape not just what people say in public but what they privately desire. High's (2012) ethnographic work in Laos addresses this *phenomenology* of development desire. Drawing on Lacanian psychoanalysis, she explains why local people continue to participate in development projects even when they have seen these repeatedly fail: "the strength and depth of their disillusionment with the state cannot be understood without an understanding also of this persistent desire for what the state is thought to be able to offer" (High, 2012: 4). People do not merely accept development narratives; they *want* them.

High's analysis finds resonance across the region. In Vietnam, Harms (2016) shows how residents of Ho Chi Minh City's new urban zones adopt state narratives of 'civilised' modern housing despite displacement and disappointment. In Cambodia, Hughes and colleagues describe villagers embracing narratives about 'good governance' and 'development'; when projects fail, blame is placed on corrupt officials rather than on the development model itself (Hughes, 2003, 2009; Hughes and Un, 2011). In northern Laos, villagers weave state narratives of prosperity into their own hopes for mobility and education, even as actual schemes deepen indebtedness (High, 2012; Turner, 2013). Roads that were supposed to bring prosperity brought instead land speculation; rubber plantations generated indebtedness; resettlement schemes provided marginalisation. Yet the narrative structure persists—because it constitutes the framework through which people understand their situation and imagine alternatives.

The depth of this penetration helps explain why critique fails to gain traction. The problem is not that affected communities lack information about dam impacts. The problem is that such information must compete with narratives that have colonised desire itself. Counter-narratives face not merely informational but affective disadvantage.

Dominant narratives as obfuscation

If narratives constitute the spaces within which development proceeds, they do so partly by obscuring what is actually happening. Shanahan et al. (2025) develop the notion of 'narrative power' drawing on Lukes' (2005) three dimensions of power, highlighting how policy actors can use narratives to shape beliefs so that people comply willingly, even against their own interests. Lukes' third dimension—the capacity to shape preferences such that dominated groups cannot articulate their interests—operates substantially through narrative. In Mekong hydropolitics, the framing of dam development as technical necessity exemplifies this: the question of whether to build recedes, leaving only questions of how. Obfuscation hides what is happening; erasure eliminates what was there before. Both are instruments of dominant narration, operating together to reshape not only perception but history itself (see below).

For Boswell (2013, 2022), narratives must be enacted by embedded political actors. Dominant narratives are consciously mobilised to manipulate beliefs and reinforce hierarchies. Institutional arrangements ensure that alternative perspectives are sidelined—not through overt censorship but through the structuring of debate and the delimitation of what counts as relevant knowledge. Blum et al. (2025) argue that power in narrative politics runs through who narrates, what is said, through which forums, and to which audiences—distinguishing 'power over' (structuring how issues are thought about), 'power to' (enabling actors to pursue interests), and 'power with' (building collective agency).

Return to Molle's nirvana concepts, for here the obscuring function becomes most apparent. Nirvana concepts work as "the narrative equivalent of looking at the hand of a magician—so that the observer does not notice that the rabbit has disappeared" (Molle, 2008: 135). 'Sustainable development' directs attention toward sustainability while development proceeds unsustainably; 'win-win cooperation' directs attention toward mutual benefit while asymmetric gains accumulate. Crucially, "narratives, just like policies, are often self-validating because they tend to produce evidence rather than the other way around. If they are at fault it is because the prescription has been either inadequately or insufficiently implemented" (Molle, 2008: 137). Each failure generates prescription for more of the same.

McGee's (1967, 1979) theory of conservation-dissolution illuminates how this works institutionally. States work to dissolve local-level power structures just enough that they can no longer resist capitalist development, while partially conserving elements of these structures when doing so helps maintain order. Nyíri's (2006) work on ethnic minorities in Yunnan illustrates the pattern: the Chinese state simultaneously celebrates minority cultures and domesticates them—rendering minorities as the "people who never stopped singing and dancing" (p. 24), their distinctiveness preserved insofar as it is emptied of political content.

Dissolution can—and often does—result in erasure: not merely the weakening of alternative structures but their elimination from the political field. And conservation can serve erasure too—when cultural forms are retained but emptied of political content, what remains is legible to the state but incapable of challenge. The dominant narrative can, as a consequence, operate through selective erasure: eliminating what threatens, preserving what can be domesticated. The temporal frame matters here. From a synchronic perspective, dominant narratives obfuscate—there are things happening that they make hard to see. From a diachronic perspective, they erase—actively eliminating what was there before. Obfuscation implies co-presence; erasure implies elimination.

The MRC exemplifies conservation-dissolution. Its mission statement—promoting "sustainable management and development... for the countries' mutual benefit and the people's well-being"—is a strong nirvana concept. When mainstream dam designs pass through the MRC's PNPCA, they emerge bearing these hallmarks. But the MRC has been dissolved in McGee's sense: stripped of the capacity to meaningfully constrain member state action. Its member states have consistently chipped away at its mandate—removing rights to halt development deemed to have transboundary impacts, focusing exclusively on the mainstream and not tributaries, 'riparianising' in ways that curtailed independence (Yong, this volume). Its form has been maintained precisely because the content has been emptied. The MRC has been conserved in a form that provides legitimation without constraint (Geheb et al., 2015).

Throughout the region, environmental safeguards have been dissolved and conserved. The elaborate framework developed around Nam Theun 2 has been conserved as exceptional: the standard that "all the developers say is not possible" to replicate (Jusi, 2011: 256). The policy framework exists in legislation but has been dissolved of practical effect for other dams built with far less scrutiny.

Infrastructure as argument

Important to hegemonic narration is the construction of infrastructure as beyond function—not merely generators but the triumph of state, "technological shrines [that] exemplify the messy dialectics between creation and destruction inherent in Modernity's Promethean Project" (Kaika, 2007: 277). As Bridge (2011: 307) notes, "Governments seeking short cuts to modernity plough resources into nuclear power, large-scale hydro, and rural electrification as a means of spreading the light of development across a territory, drawing citizens out of the metaphorical darkness of tradition".

Consider China's Nuozhadu dam: "a testament to China's ambitious commitment to infrastructure development and… a pivotal project to the Western Development strategy, '12th 5-Year Plan,' 'West-to-East Power Transmission,' and 'Yunnan Electricity Export Strategy'" (Liu et al., 2025: 135). This rockfill dam is the tallest of its type in China, and the third tallest in the world; it has the largest spillway in Asia, the largest underground powerhouse in China, and the largest hydropower dam in the Mekong basin. These superlatives are not technically significant for electricity generation. They are assertions of dominion—claims about state capacity directed at domestic populations, ethnic minorities, neighbouring states, and an international community. The dam is the argument. And it is no coincidence that the Mekong Region's dam sites cluster in places characterised as wild, backward, and under-regulated—the highlands of Zomia (van Schendel, 2002), where state authority has historically been weakest. The dam answers illegibility with concrete.

Large hydraulic infrastructure consists of "materially tying together the country and physically making the presence of an irresistible Leviathan felt" (Mohamud and Verhoeven, 2016: 183). Infrastructure and ideology are mutually constituting: "Infrastructure as text; economy as pretext; politics as subtext" (Appel, 2018: 48).

Rupturing connectivity

Borrowing from Mahanty et al. (2023), we can describe the costs of Mekong development as the severing of connectivity—hydrological, ecological, and social—that historically integrated the basin. These ruptures are not unfortunate byproducts but the constitutive costs of the infrastructural narrative.

Water itself is the most obvious ruptured flow. China now has the capacity to fundamentally alter downstream hydrology through the Lancang cascade. The seasonal flood pulse that historically drove the Mekong's productivity—swelling Cambodia's Tonle Sap Lake to six times its dry-season size, triggering fish migrations—is increasingly disrupted by upstream storage decisions. Hecht et al. (2019) estimate that water storage has increased from 2 per cent of annual flow in 2008 to potentially 20 per cent by 2025.

Sediment represents a second rupture. The Mekong's sediment load has already declined by approximately 70 per cent; if all planned dams are built, sediment reaching the delta could fall by 97 per cent (MRC, 2018). The consequences are severe: reduced floodplain fertility, accelerated erosion, delta subsidence, and saltwater intrusion in Vietnam's most productive agricultural region. Rousseau and Dinh (this volume) show how sand has only recently emerged as a concern in MRC discourse—a case study of what goes unseen when particular framings dominate.

Fish constitute a third rupture. The Mekong supports one of the largest freshwater fisheries in the world, providing protein for tens of millions. Large numbers of species undertake seasonal migrations now blocked by dams. The MRC's projections suggest that fish biomass could decline by 40 to 80 per cent by 2040, with losses potentially reaching USD 23 billion. Sithirith and Geheb (this volume) apply state capture analysis to Cambodia's Tonle Sap, showing how chronic mismanagement reflects the capture of governance by elite interests.

Information is a fourth rupture. Baird (this volume) notes that information about negative impacts "is often repressed through various means, such as by limiting democratic rights and freedoms, including free speech and access to information; and also by controlling and influencing the media". In Laos, open opposition to dams is not permitted. In Cambodia, civil society space has severely contracted.

Perhaps the most significant rupture is that of voice. Indigenous and local knowledge is marginalised through multiple mechanisms. Silvano et al. (2023) demonstrate how fishers' multidimensional knowledge could advance fisheries science but is routinely overlooked. When data contradict development agendas, they are "insufficiently used to affect hydropower development policy, because they do not support the agendas of economically and politically powerful dam developers and their allies" (Baird, this volume).

Manorom and Geheb (this volume) develop an analysis distinguishing between voice-as-power—the state's voice that acts upon the world—and voice-as-performance—villagers' voices that are merely noted and filed. The MRC's PNPCA, presented as pluralistic consultation, actually functions as a mechanism for projecting unified state messaging. Käkönen and Hirsch (2009) identified the 'anti-politics' positioning of Mekong knowledge production: the ways in which political aspects of decision making are denied and projects justified as technically logical. Green and Baird (2020) show how environmental assessment procedures produce scale—determining what counts as an 'impact area' and who qualifies as 'project-affected'—in ways that serve developer interests.

Bright et al. (this volume) offer an alternative vision through the Salween Peace Park. The *kaw* system—Karen indigenous governance integrating land, water, spirits, and usage rights—embodies a different relationship between knowledge, authority, and landscape. The Peace Park represents 'counter-territorialisation': the production of territory by communities rather than states.

Where to and what now?

The title of this chapter—'The Bliss of Development'—is drawn from a 2023 Xinhua article describing how "BRI hydropower projects send bliss of development to Laos". The phrase crystallises what this volume seeks to contest. It is a narrative claim masquerading as a description, a hegemonic assertion presenting itself as common sense. The 'bliss' exists—but for whom? Certainly for the political and economic elites who profit from construction contracts, for the politicians who receive rents, and for the consumers whose air conditioners run on Mekong electricity but less so for fishing communities whose catches are declining, for the farmers whose soil fertility is diminishing, or for the delta dwellers facing saltwater intrusion.

The volume's empirical chapters substantiate these claims across the mainland Southeast Asian hydropolitical space. Zhang et al. reveal the complexity of 'China' as an actor. Sopera exposes the 'hydrocorruption' that drives Lao dam building. Sithirith and Geheb demonstrate state capture in Tonle Sap governance. Rousseau and Dinh show what goes unseen when sediment finally enters MRC discourse. Tran and Tuan trace 'forced adaptation' imposed on Vietnamese delta farmers. Grundy-Warr develops 'biophysical geopolitics' through attention to rocks, dolphins, and flood pulses. Manorom and Geheb propose 'deep listening' as methodology and normative aspiration. Bright et al. document the Salween Peace Park as counter-territorialisation. Baird synthesises these concerns through careful attention to how flows are facilitated and ruptured through political choices presented as technical necessity. And Yong traces the MRC's institutional trajectory toward conservation-dissolution.

The ecological evidence does not permit optimism. The Stimson Center's Mekong Dam Monitor (2024) documents ongoing impacts from upstream operations. The MRC's own scenarios project severe declines in sediment and fisheries by 2040. Climate change compounds these pressures. The political economy that drives dam building—the hydrocorruption that Sopera analyses and the circuits of power that Marks and Zhang (2019) trace—shows no signs of abating. The narrative space remains largely closed.

Yet recognition is itself a form of power. As Baird (this volume) notes, the first step toward finding appropriate solutions is "recognising the nature of present and future problems, and their links to an array of different types of hydropolitics". By revealing how hydropolitical assemblages are produced and operate, by tracing the flows and ruptures that constitute the contemporary Mekong, and by attending to those whose voices are amplified and those whose voices are silenced, we can begin to imagine different trajectories. The countries sharing the Mekong Basin would need to move, as Baird puts it, "from politics-impeding sustainability discourses, and more towards rights-based approaches". This is a tall order in a region where authoritarian infrastructuralism remains dominant. But the ecological and social costs of the present trajectory are mounting. Perhaps far closer than the horizon now is the emerging second contradiction of capital.

References

Appel, H. 2018. Infrastructural time. In Anand, N., Gupta, A. and Appel, H. (eds.), *The promise of infrastructure*, 42–61. Durham (NC): Duke University Press.

Bakker, K. 1999. The politics of hydropower: developing the Mekong. *Political Geography* 18(2): 209–232, https://doi.org/10.1016/S0962-6298(98)00085-7.

Blum, S., Jones, M.D., Kuenzler, J., Kuhlmann, J., Parth, A.-M. and Vogeler, C.S. 2025. Conceptualizing dimensions of power in the narrative policy framework: how does 'who says what to whom' matter? *SSRN* June 18, 2025, http://dx.doi.org/10.2139/ssrn.5316103.

Boswell, J. 2013. Why and how narrative matters in deliberative systems. *Political Studies* 61(3): 620–636, https://doi.org/10.1111/j.1467-9248.2012.00987.x.

Boswell, J. 2022. Narrative analysis. In Ercan, S.A., Hendriks, C.M., Dryzek, J.S. and Boswell, J. (eds.), *Research methods in deliberative democracy*, 310–323. Oxford: Oxford University Press.

Bridge, G. 2011. Past peak oil: political economy of energy crises. In Peet, R., Robbins, P. and Watts, M. (eds.), *Global political ecology*, 307–324. Abingdon: Routledge.

Dwyer, M.B. 2020. "They will not automatically benefit": the politics of infrastructure development in Laos's Northern Economic Corridor. *Political Geography* 78(2020): 102118, https://doi.org/10.1016/j.polgeo.2019.102118.

Forsyth, T. 2003. *Critical political ecology: the politics of environmental science*. London: Routledge.

Geheb, K., West, N. and Matthews, N. 2015. The invisible dam: hydropower and its narration in the Lao People's Democratic Republic. In Matthews, N. and Geheb, K. (eds.), *Hydropower development in the Mekong Region: political, socio-economic and environmental perspectives*, 101–126. London: Earthscan.

Gramsci, A. 1971. *Selections from the prison notebooks*. Edited and translated by Q. Hoare and G. Nowell Smith. London: Lawrence and Wishart.

Green, W.N. and Baird, I.G. 2020. The contentious politics of hydropower dam impact assessments in the Mekong River basin. *Political Geography* 83(2020): 102272, https://doi.org/10.1016/j.polgeo.2020.102272.

Hajer, M. 1995. *The politics of environmental discourse: ecological modernisation and the policy process*. Oxford: Clarendon Press.

Harms, E. 2016. *Luxury and rubble: civility and dispossession in the new Saigon*. Berkeley: University of California Press.

Harvey, D. 1981. The spatial fix — Hegel, Von Thünen, and Marx. *Antipode* 13(3): 1–12, https://doi.org/10.1111/j.1467-8330.1981.tb00312.x.

Harvey, D. 1993. The nature of environment: the dialectics of social and environmental change. *Socialist Register* 29(1993): 1–51, https://socialistregister.com/index.php/srv/article/view/5621.

Harvey, D. 2014. *Seventeen contradictions and the end of capitalism*. London: Profile Books.

Hecht, J.S., Lacombe, G., Arias, M.E., Dang, T.D. and Piman, T. 2019. Hydropower dams of the Mekong River basin: a review of their hydrological impacts. *Journal of Hydrology* 568(2019): 285–300, https://doi.org/10.1016/j.jhydrol.2018.10.045.

High, H. 2012. *Fields of desire: poverty and policy in Laos*. Singapore: NUS Press.

Hughes, C. 2003. *The political economy of Cambodia's transition, 1991–2001*. London: RoutledgeCurzon.

Hughes, C. 2009. Cambodia in 2008: consolidation in the midst of crisis. *Asian Survey* 49(1): 206–212, https://doi.org/10.1525/as.2009.49.1.206.

Hughes, C. and Un, K. 2011. *Cambodia's economic transformation*. Copenhagen: NIAS Press.

Jusi, S. 2011. Challenges in developing sustainable hydropower in Lao PDR. *International Journal of Development Issues* 10(3): 251-267, https://doi.org/10.1108/14468951111165377.

Kaika, M. 2007. Dams as symbols of modernization: the urbanization of nature between geographical imagination and materiality. *Annals of the Association of American Geographers* 96(2): 276–301, https://doi.org/10.1111/j.1467-8306.2006.00478.x.

Käkönen, M. and Hirsch, P. 2009. The anti-politics of Mekong knowledge production. In Molle, F., Foran, T. and Käkönen, M. (eds.), *Contested waterscapes in the Mekong Region: hydropower, livelihoods and governance*, 333–365. London: Earthscan.

Linton, J. and Budds, J. 2014. The hydrosocial cycle: defining and mobilizing a relational-dialectical approach to water. *Geoforum* 57(2014): 170–180, https://doi.org/10.1016/j.geoforum.2013.10.008.

Liu, B., Zhang, Z., Yan, L., Liu, C., Wu, M., Yu, Y. and Zhang, D. 2025. Comprehensive review of innovative construction technology of a high-core rockfill dam: Nuozhadu project. *River* 4(2): 135–148, https://doi.org/10.1002/rvr2.70004.

Lukes, S. 2005. *Power: a radical view*. 2nd ed. Basingstoke: Palgrave Macmillan.

Magee, D. 2006. Powershed politics: Yunnan hydropower under Great Western Development. The *China Quarterly* 185(2006): 23–41, https://doi.org/10.1017/S0305741006000038.

Mahanty, S., Milne, S., Barney, K., Dressler, W., Hirsch, P. and To, P.X. 2023. Rupture: towards an emplaced, critical, and experiential view of nature-society crisis. *Dialogues in Human Geography* 13(2): 177–196, https://doi.org/10.1177/20438206221138057.

Marks, D. and Zhang, J. 2019. Circuits of power: environmental injustice from Bangkok's shopping malls to Laos' hydropower dams. *Asia Pacific Viewpoint* 60(3): 296–309, https://doi.org/10.1111/apv.12242.

Massey, D. 1993. Power-geometry and a progressive sense of place. In Bird, J., Curtis, B., Putnam, T. and Tickner, L. (eds.), *Mapping the futures: local cultures, global change*, 60–70. London: Routledge.

Massey, D. 1994. *Space, place, and gender*. Minneapolis (MN): University of Minnesota Press.

Massey, D. 2012. *For space*. London: Sage.

McGee, T. 1967. *The Southeast Asian city: a social geography of the primate cities of Southeast Asia*. New York (NY): Praeger.

McGee, T. 1979. Conservation and dissolution in the Third World city: the 'shanty town' as an element of conservation. *Development and Change* 10(1): 1–22, https://doi.org/10.1111/j.1467-7660.1979.tb00028.x.

Middleton, C. and Allouche, J. 2016. Watershed or powershed? A critical hydropolitics of the Lancang-Mekong Cooperation Framework. *The International Spectator* 51(3): 100–117, https://doi.org/10.1080/03932729.2016.1209385.

Mohamud, M. and Verhoeven, H. 2016. Re-engineering the state, awakening the nation: dams, Islamist modernity and nationalist politics in Sudan. *Water Alternatives* 9(2): 182–202, http://www.water-alternatives.org/index.php/alldoc/articles/312-a9-2-2/file.

Molle, F. 2008. Nirvana concepts, narratives and policy models: insights from the water sector. *Water Alternatives* 1(1): 131–156, https://www.water-alternatives.org/index.php/alldoc/articles/vol1/v1issue1/20-a-1-1-8/file.

MRC (Mekong River Commission). 2018. The Council Study: study on the sustainable management and development of the Mekong River Basin. Vientiane: MRC.

Nyíri, P. 2006. *Scenic spots: Chinese tourism, the state, and cultural authority*. Seattle (WA): University of Washington Press.

Riedke, E. and Rottenburg, R. 2016. Introduction. In Riedke, E. (ed.), *The ordering power of narratives*. Working Papers of the Priority Programme 1448 of the German Research Foundation, No. 21, 5–15. Leipzig and Halle: DFG Priority Programme 1448, https://lost-research-group.org/wp-content/uploads/2017/05/SPP1448_WP21_Riedke.pdf (accessed 6 January 2026).

Robbins, P. 2012. *Political ecology: a critical introduction*. 2nd ed. Chichester: Wiley-Blackwell.

Roe, E. 1991. Development narratives, or making the best of blueprint development. *World Development* 19(4): 287–300, https://doi.org/10.1016/0305-750X(91)90177-J.

Roe, E. 1994. *Narrative policy analysis: theory and practice*. Durham (NC): Duke University Press.

Roe, E. 1999. *Except-Africa: remaking development, rethinking power*. New Brunswick (NJ): Transaction Publishers.

Shanahan, E.A., Jones, M.D., McBeth, M.K., Radaelli, C.M., Tosun, J. and Weible, C.M. 2025. Narrative power in the narrative policy framework. *Policy Studies Journal* 26 May 2025, https://doi.org/10.1111/psj.70038.

Silvano, R.A.M., Baird, I.G., Begossi, A., Hallwass, G., Huntington, H.P., Lopes, P.F.M., Parlee, B. and Berkes, F. 2023. Fishers' multidimensional knowledge advances in fisheries and aquatic science. *Trends in Ecology and Evolution* 38(1): 8–12, https://doi.org/10.1016/j.tree.2022.10.[illegible]

Sneddon, C. and Fox, C. 2006. Rethinking transboundary waters: a critical hydropolitics of the Mekong basin. *Political Geography* 25(2): 181-202, https://doi.org/10.1016/j.polgeo.2005.11.002.

Stimson Center. 2024. Mekong Dam Monitor Annual Report: 2023–2024. Washington, DC: Stimson Center, https://www.stimson.org/2024/mekong-dam-monitor-annual-report-2023-2024/ (accessed 6 January 2025).

Suhardiman, D., Giordano, M. and Molle, F. 2012. Scalar disconnect: the logic of transboundary water governance in the Mekong. *Society and Natural Resources* 25(6): 572–586, https://doi.org/10.1080/08941920.2011.604398.

Swyngedouw, E. 2004. *Social power and the urbanization of water: flows of power*. Oxford: Oxford University Press.

Turner, S. 2013. Under the state's gaze: upland trading-scapes on the Sino-Vietnamese border. *Singapore Journal of Tropical Geography* 34(1): 9–24, https://doi.org/10.1111/sjtg.12010.

van Schendel, W. 2002. Geographies of knowing, geographies of ignorance: jumping scale in Southeast Asia. *Environment and Planning D: Society and Space* 20(6): 647–668, https://doi.org/10.1068/d16s.

Wong, T. 2010. Making the Mekong: nature, region, postcoloniality. PhD Thesis. Columbus (OH): Department of Geography, Ohio State University, http://rave.ohiolink.edu/etdc/view?acc_num=osu1275482743 (accessed 5 January 2026).

Xinhua. 2023. Asia Album: BRI hydropower projects send bliss of development to Laos. *Xinhuanet* August 3 2023, https://english.news.cn/20230803/f2ca1307ac0d40ed95218772b03c5ca3/c.html (accessed 5 January 2026).

CHAPTER 2

Hydropolitics in the Mekong: The role of Chinese subnational governments

Hongzhou Zhang,[1] Kim Geheb[2] and Genevieve Donnellon-May[3]

Abstract

The Mekong River houses the largest inland fishery in the world and offers food security and a means of subsistence to millions of people. Its basin has, however, emerged as one of the most water conflict-prone regions in recent years. One of the main triggers for water tensions in the Mekong is China's actions to utilise these shared water resources, particularly its dam-building activities. Since the 1980s, mega dams have been built on the upstream of the river. With the completion of the cascade of dams in Yunnan, the focus of dam development in China has been moving northwards to Tibet. While studies are abundant on the role played by various national actors in Mekong hydropolitics, little research has been conducted at the subnational level, particularly in the context of China. Subnational governments have increasingly constituted a vital link in the multi-level governance of water in the basin. This chapter intends to fill this gap through a comparative study of the roles played by two Chinese provinces—Yunnan and Tibet—in shaping the hydropolitics of the Mekong.

Keywords: Mekong River Basin, hydropolitics, China, sub-national government, Yunnan Province, Tibet Autonomous Region

Introduction

The Mekong River is one of the most biodiverse and geopolitically important river systems in the world. As the bedrock of the world's largest inland fishery, the Mekong River is tied to the food security and livelihood of millions of people across the nations that share its basin. In recent decades, the Mekong River Basin (MRB) has become a flashpoint for escalating water conflicts, triggered by dam-building activities by upper riparian states, including China (Biba, 2018; Zhang and Li, 2020; Zhang and Zhang, 2021). Since the 1980s, when China began its dam construction

1 S. Rajaratnam School of International Studies, Nanyang Technological University, Singapore.
2 Landscape Alliance, Nairobi, Kenya.
3 Asia Society Policy Institute and the Asia Society Australia.

on the Mekong (called the Lancang in China), the country has faced significant criticism for the environmental and ecological damage that its dams have caused (Zhong et al., 2016). Over the years, these Chinese mega hydro-engineering projects have become one of the main sources of conflict between China and downstream Southeast Asian countries. These disputes are further exacerbated by China's reluctance to join the Mekong River Commission (MRC) (Biba, 2018; Chen et al., 2013; Ho, 2014, 2017; Pohl et al., 2014; Yeophantong, 2017; Zhang and Li, 2017, 2018).

The Mekong River rises in southeastern Qinghai Province, then flows through the Tibetan Autonomous Region and then southeast into Yunnan Province, before entering Southeast Asia. During its passage through Chinese territory, the river typically follows steep-sided V-shaped valleys, and descends 4,000 metres, making it ideal for large-scale, reservoir hydropower production. It is one of the most important sources of hydroelectricity for the country (Brauer and Kliem, 2017). Since the 1980s, China's unilateral dam building on the Lancang has triggered concerns and criticism from lower Mekong River countries and the international community (Zhong et al., 2016). Various diplomats and security experts have begun to describe the Mekong River as 'the next South China Sea', given the potential of the river to spark regional tensions and derail relations between China and its Southeast Asian neighbours (Zhang and Li, 2017).

In this chapter, hydropolitics refers to the interaction between various actors that directly and indirectly shape the allocation and/or use of shared water resources. The existing body of scholarly work on Mekong hydropolitics has been prolific in analysing the influence of national governments and international institutions, including river basin organisations, regional institutions, and cooperation frameworks (Busbarat et al., 2021; Gerlak and Haefner, 2017; Goh, 2004; Grumbine et al., 2012; Han, 2017; Middleton and Devlaeminck, 2021; Spijkers and Devlaeminck, 2022). There is, however, limited research in the current literature concerning the role of subnational actors—such as provincial governments—especially within China's complex governance structures. These are often depicted as mere extensions of the central government, attributing them minimal policy autonomy. This portrayal overlooks the potential distinctiveness and influence of these subnational actors within the broader political and environmental dynamics of the Mekong Region.

In recent years, there have been some attempts to study the role of Yunnan in China's overall foreign policy and Mekong water policy (Li, 2014; Zhang and Li, 2020). Nonetheless, one overlooked actor is the Tibet Autonomous Region (TAR) government. The absence of research on the role of the TAR government is problematic, as with the completion of the cascade of dams in Yunnan, the focus of dam development in China has been moving northwards from Yunnan to Tibet. It is important to recognise that provincial governments, such as those of Yunnan and Tibet, are not passive implementers of national policies but active agents shaping the course of water governance in the MRB (Liu and Song, 2019; Zhang and Li, 2020). Notably, provincial governments engage in policy formulation and implementation, project execution, and even diplomatic negotiations with other riparian states. Doing so enables them to shape regional water security, economic development, and geopolitical relations.

This chapter provides a comparative study of the roles played by Yunnan and Tibet in the unfolding scenario of Mekong hydropolitics. Through this lens, we aim to expand the academic discourse beyond the traditional focus on national and international actors, thereby offering a more nuanced understanding of multi-level governance in transboundary water management. The impetus for this study is not merely academic; it holds significant policy implications. By understanding the role of these subnational entities, policymakers can more effectively negotiate, strategise, and collaborate to foster a sustainable and equitable utilisation of the Mekong's resources.

Yunnan's role in shaping Mekong Hydropolitics

China's approach towards Mekong has shifted notably since 2015. Previously, the approach attempted to balance unilateral development imperatives and selective multilateral cooperation (Guo, 2017; Ho, 2014). Since 2015, however, it has adopted an increasingly active multilateral approach via the Lancang Mekong Cooperation (LMC) mechanism. While Beijing dominates this strategic calculation, Yunnan has contributed to shaping it in a variety of ways: (a) influencing China's overall relations with Southeast Asian countries; (b) direct involvement or investment of Yunnan-registered firms in LMB infrastructure; and (c) domestic development of the water resources of the Lancang River (Mekong River) within its jurisdiction.

The province plays an important role in shaping China's relations with Southeast Asian countries (Liu and Song, 2019; Tubilewicz, 2017). This role has evolved significantly since the early 1990s as the province has increasingly augmented its 'bridging' role between China and Southeast Asia. Initially, Yunnan played a significant role in the Greater Mekong Sub-region (GMS), an economic cooperation platform led by the Asian Development Bank which it used to launch a variety of local and multilateral engagements (Ren and Li, 2018). By the early 2000s, Yunnan had secured support from Beijing to further enhance its bridging role. This manifested in the planning and execution of infrastructure projects that helped to augment the province's connectivity and economic integration with the region. The Yunnan-Myanmar oil and gas pipeline, for example, was initiated by the Yunnan Provincial Government (YPG). This project not only reflected the province's strategic activism but also its leveraging of broader Chinese concerns by capitalising on national energy security apprehensions (Wong, 2018). Initiatives such as this, as well as other connectivity through infrastructural development, has led Beijing to recognise Yunnan's 'Gateway' initiative as a national strategy (Li, 2014).

The unveiling of China's Belt and Road Initiative (BRI) in 2013 by President Xi Jinping provided Yunnan additional opportunities. Its launch was a great opportunity for Yunnan, which had long felt that the GMS needed an 'upgrade'. The BRI's focus on infrastructure development, trade, and investment dovetailed perfectly with Yunnan's long-standing objectives. Yunnan moved robustly to engage with the BRI apparatus, and to demonstrate the province's willingness to utilise the BRI as a conduit for fulfilling its Southeast Asian Gateway ambitions. It has created a Yunnan Liaison Office of the LMC Secretariat, which is the first and only such office set up at the provincial level, and a suite of cooperation centres offering training and other capacity building around law enforcement, the environment, and higher education. The province proactively tries to support the LMC by hosting high-level meetings amongst the partner countries. Scholars from Yunnan University produce an annual edition of the 'Lancang Mekong Cooperation Blue Book' which has become an important resource for Chinese scholars and policymakers to understand the annual politics, economy, diplomacy, and societal development trends in the LMC partner countries and key issues of subregional cooperation.

Spinning off these initiatives, the YPG has sought to develop relationships with lower Mekong governments through a wide variety of formal visits by YPG officials to lower Mekong governments, the establishment of working groups, or through the creation of 'meeting mechanisms'. To coordinate all of this activity, the YPG has established a coordinating office for LMC activities, cementing its role as a key facilitator within the cooperation framework.

The above diplomatic initiatives have served both to advance Beijing's political and cooperative agenda into the lower Mekong, as well as Yunnan's own. Insofar as Yunnan's engagement in Mekong-Lancang hydropolitics is concerned, there have been two key interest groups within Yunnan. The first of these is focused on hydropower development (the 'dam coalition'), while the second is focused on commercial navigation (the 'transport coalition'), both of which have worked to lobby the YPG—and the central government—to represent their respective interests.

The dam coalition has been spearheaded by the Yunnan Provincial Energy Department, part of the YPG. It includes within it very powerful power companies. Most of China's five giant power firms (Huadian, PowerChina, Datang Corp., Huaneng Group, and China Energy Investment Group (CEIC)), have large offices in Kunming to take advantage of the substantial hydropower opportunities that Yunnan represents. HydroLancang holds the concession for the construction and operation of all dams on the Lancang mainstream, placing it in a highly strategic position with respect to the lower Mekong countries. Huadian (now merged into CEIC) holds the concession on the Nu Jiang (Salween) hydropower dams (if these commence) through Nu Jiang Hydropower Company; on the Jinsha, PowerChina through its Yunnan-based subsidiaries, has been heavily involved in hydropower construction. The dam coalition has significant commercial interests south of the Yunnan border, which in turn has hydropolitical implications. China is, generally, a significant purveyor of global hydropower (McDonald et al., 2009). Many of its dams hold back very significant volumes of water—combined, the existing 12 mainstream dams along the Lancang mainstream in Yunnan have a total storage capacity of 43.2 km^3—some 9 per cent of the total average annual Mekong discharge (calculated from MERFI, 2025). Yunnan's hydropower dams have had a demonstrable effect on the Mekong's sediment load, and on its hydrological rhythms (e.g., Hecht et al., 2019; Kondolf et al., 2018), which in turn has implications for the Mekong's aquatic ecology, agriculture, fisheries, and other dependent systems. This highly strategic position has been addressed elsewhere in the literature (e.g., Biba, 2012, 2018; Han, 2017; Wang et al., 2022; Zhang and Li, 2020).

The 'transport coalition' is led by the Yunnan Provincial Transportation Department (YPTD), consisting of shipping merchants and commercial traders. Since the late 1980s, the province has aggressively pursued commercial navigation, most notably with the initiation of the Lancang-Mekong River international navigation channel in 1989 (Qiao, 1996). From supporting river navigation research and direct subnational interaction with other Mekong countries to funding major river clearance programmes and pilot voyages both within Yunnan and in other Mekong River countries as well as lobbying for the central government's policy and financial support, Yunnan has played a significant role in the efforts to develop international navigation along the Lancang-Mekong River. Yunnan's efforts led to the eventual signing of the Quadripartite (China, Myanmar, Laos, and Thailand) Agreement on Commercial Navigation on the Lancang-Mekong River in 2000. Considering that international commercial navigation on the Mekong cannot be unilateral, the 'navigation coalition' in Yunnan saw multilateral cooperation as necessary for providing Yunnan with access to the sea via the Mekong River. This clashes with the interests of the 'dam coalition', whose political influence grew due to the importance of hydropower for national development and Yunnan. China has the world's largest exploitable hydropower resources, mainly in the southwest, central, and northwest, while major energy consumption areas are in the east and south. Thus, a 'gigantic west-to-east hydropower electricity transfer' is a key national strategy. Yunnan, despite being impoverished, has the second-highest hydro-energy potential in China and aims to become an energy hub for China and Southeast Asia, hoping dam construction will boost employment and economic growth.

Nevertheless, under the BRI, there has been renewed interest and greater enthusiasm from the navigation coalition to turn the Lancang-Mekong River into a 'golden waterway' for Yunnan. Yunnan took advantage of the LMC to revive the 'Development Plan for International Navigation on the Lancang-Mekong River (2015–2025)' and succeeded in obtaining approval for it from the Thai authorities in December 2016. Following a 13-year suspension of the shipping channel, two freight vessels loaded with sand left Simao on the Lancang River for Jinghong in June 2018, marking the resumption of cargo services. Between 2019 and 2020, China (with funding from both national and Yunnan governments) spent over 1 billion yuan in constructing a Grade IV Waterway (which can reliably handle 500-tonne vessels) in the Lancang River (YPTD, 2022). In 2022, the YPTD issued the '14th Five-Year Plan for Waterway Transportation in Yunnan Province', one of the key priorities of which is to promote the construction of the Lancang-Mekong international corridor to serve Yunnan's opening-up strategy (YPTD, 2022). In addition to an estimated increase in cargo volume of the Lancang Mekong River, which could reach 3 million tonnes by 2025, the upgrade of the Lancang Mekong waterways is also expected to significantly boost Yunnan's tourism industry (YPTD, 2022).

Meanwhile, there are new developments with regard to the navigation coalition. First, with near completion of the dams on the Lancang downstream in Yunnan, the focus is moving to Tibet. Second, as hydropower development in Yunnan reached its limit (i.e., very few opportunities for new dams), hydropower companies in Yunnan and other provinces are expanding into Southeast Asia (in addition to moving to Tibet; see the following section). Southeast Asia receives over half of China's global hydropower investments. As such, stable and friendly ties between China and other Mekong River countries, including in relation to transboundary water issues, are becoming important for these hydropower companies as well.

For China's power companies, Yunnan is an effective springboard from which to penetrate the Southeast Asian hydropower and related electricity markets—supported by the YPG's diplomatic efforts, channelled through the LMC. Thus, HydroLancang both built and operates Cambodia's largest hydropower dam, the Lower Sesan 2, as well as the Shweli 1 dam in northern Myanmar; SinoHydro, a subsidiary of PowerChina, built and operated the seven-dam Nam Ou cascade and northeastern Laos. In 2022, China Southern Power Grid (CSG) Yunnan International Co. Ltd., a subsidiary of China Southern Power Grid, and Électricité du Lao (EdL) "successfully achieved two-way transmission" between the two countries for the first time since the 115 kV transmission line was built in 2010. Yunnan will purchase excess Lao-produced power between June and October, when Lao power is abundant, and sell Laos power during the dry season (The Laotian Times, 2022). The power will come from the Nam Tha 1 hydropower plant, a 168 MW dam itself operated by CSG International, constructed by PowerChina, and financed with a loan from China EXIM Bank. CSG-Yunnan International is also the majority shareholder (believed to hold some 90 per cent of the shares) in Laos' national electricity grid firm. This joint venture was launched in 2024.

The convergence of interests of these two coalitions has, to a certain extent, led the YPG to lobby for the LMC and to actively participate in various LMC projects related to transboundary water cooperation. One of the most striking aspects of Yunnan's influence is its dominance in the academic discourse surrounding the Mekong River. Yunnan University has been a leading institution in Mekong-related research. This academic prominence provides Yunnan with a substantial platform to influence discourse and policy directions, a phenomenon further evinced by the thematic focus of its research output. The latter predominantly concentrates on Yunnan's specific interests in the river basin, particularly in the areas of hydropower development and commercial navigation. It is thus hardly surprising that Yunnan scholars have pioneered research on water diplomacy in China, under the leadership of Zhang Li, Lu Guangsheng, He Daming, and a few other researchers from Yunnan University. Before the BRI was officially proposed, research by Zhang and Lu centred on advocating GMS upgrading and broadening to include transboundary water governance; after the LMC, their research has focused on the rationale and method of adopting water diplomacy under the LMC. In addition to national interests, Yunnan's own interests are key considerations too. For instance, Zhang and Lu (2016) suggest adopting water diplomacy in the Mekong River to accelerate Yunnan's own economic growth. From the perspective of Chinese scholars and officials, adopting water diplomacy can help Yunnan Province to create opportunities for its hydropower sector, river shipping, and cross-border trade.

With its responsibilities for foreign relations, Beijing dominates intra-regional Chinese relations with the lower Mekong's states. Key aspects of this initiative are to play down the relevance of China's hegemonic position at the top of the basin (e.g., Tian and Liu, 2016), and the extent to which it can control the flows of the Mekong's headwaters. In the latter case, the control afforded by its dams is cast in a positive light by arguing that dry season flows are increased, while wet season flows (and floods) are reduced (e.g., Yuwei, 2022). Beijing works hard to be seen as benevolent at worst, and a willing and helpful neighbour at best. A 2010 drought that severely affected lower Mekong countries led to much finger-pointing at Chinese Mekong dams. China responded by extending its data-sharing agreement with the Mekong River Commission (MRC) to include (some) data on dry-season flows and announced that it would be cancelling the planned Mengsong hydropower plant (which would, had it been constructed, have been the southernmost dam in its cascade). There can be little doubt that Beijing took seriously the barrage of criticisms emanating from lower Mekong NGOs, CSOs, and (where permitted) articles and editorials in the press (Biba, 2016).

Four years later, another drought occurred. On this occasion, China responded to a formal request from the Vietnamese Government by ordering the release of water from its Jinghong hydropower plant (currently, the lowermost dam in its cascade). Over three phases in 2016, the dam released a total of 12.65 billion m^3 of water to thirsty lower Mekong countries (MRC and MoWRoC, 2016). These gestures were accompanied by further solidifying data-sharing agreements with the MRC. Nevertheless, China's geopolitical position on the Mekong and the intensity of its mega dam construction mean that it still attracts criticism—most recently, the suggestion that flooding in the downstream Mekong countries in September 2024 was because of releases from its dams (e.g., SCMP, 2024).

Zhong Yong, the former head of the LMC's Water Cooperation Center, complained in 2022 that

> certain countries and organization [*sic.*] out of the region deliberately ignore the facts and mislead the public by using baseless ideas which are totally out of the reality of Lancang-Mekong countries ...They always put the responsibility of flood and drought disasters on reservoirs and dams. Such behaviors deny the contribution of water infrastructure to the regional economic and social development, as well as to local livelihoods (quoted in Yuwei, 2022).

Not without merit, China sees infrastructural development (and associated financing) as a viable diplomatic instrument—which Yunnan and its construction firms can deliver, helping to support Yunnan's economy (which Beijing continues to regard as under-developed), and the infrastructural and construction ambitions of the lower Mekong countries.

Tibet as a Newcomer in Mekong Hydropolitics

Tibet's marginal influence on Mekong hydropolitics in the past

The Tibet Autonomous Region (TAR) government has been markedly less involved in Lancang-Mekong hydropolitics than the YPG. This is a consequence of a multifaceted array of economic, political, and geographical factors which have collectively circumscribed Tibet's influence in this hydropolitical arena.

The first factor is the geographical positioning of China's dam-building activities along the Lancang River. Unlike Yunnan, where the lower reaches of the Lancang River have been a focal point for hydroelectric projects since the 1980s, the TAR has been relatively bypassed in these initiatives. Yunnan's topographical endowment has seen it attract far more dam construction than the TAR reaches of the Lancang. Hence, Yunnan is more actively involved in governance and developmental projects. The absence of significant economic development activities along the Lancang River in the TAR further diminishes its motivation to engage in hydropolitics. Geographical challenges exacerbate the costs of hydropower development in the TAR—the region's high-altitude terrain and climatic conditions create an environment where the costs of infrastructure projects are considerably higher than at lower elevations.

Second, the TAR's unique economic and political status as an autonomous region within China imposes significant constraints on its external economic engagement, diplomatic endeavours, and cross-border interactions—the TAR, unlike Yunnan, has no 'bridgehead' status. Among all the provincial entities, Tibet has the country's smallest, most remote, and most sparsely populated economy. While Yunnan has vigorously pursued regional integration and economic cooperation, the TAR's interactions with neighbouring countries have been minimal and largely circumscribed. Also, the TAR is far from industrialised, dominated as it is by animal husbandry and farming. Historically, the TAR has been reliant on electricity imports from other Chinese provinces, such as Xinjiang and Qinghai, to fulfil its own comparatively modest energy needs (Zhang, 2016).

Thirdly, there is also a severe lack of scholarly and bureaucratic advocacy in Tibet. As discussed in the previous section, an important element underpinning Yunnan's active role in Lancang-Mekong hydropolitics is its prominent and influential scholarly community and capable government officials. Tibet's transboundary water research is almost non-existent. The absence of an influential academic community or a cadre of bureaucrats with expertise in transboundary water governance has resulted in an advocacy void for the TAR's role in this arena.

These geographical, economic, and political variables have collectively relegated the TAR to the periphery in the multi-level governance architecture of this critical transboundary waterway. Unlike Yunnan, which has capitalised on a combination of geographical advantages, political activism, and intellectual advocacy to assert itself as a significant actor in the hydropolitics of the region, the TAR has been largely sidelined due to these constraints. Thus, understanding these multi-dimensional factors is integral to comprehending the dynamics of subnational involvement in Lancang-Mekong water governance.

Tibet's growing hydropolitical relevance on the Lancang

In recent years, the TAR has begun to assume greater relevance in the Lancang-Mekong's hydropolitics (Chinanews, 2009). To begin with, similar to that of Yunnan and other Chinese provinces, since the launch of the BRI in 2013, the TAR government has played an increasingly proactive role in reaching out to its neighbouring provinces and countries and taking the initiative to enhance cross-border economic cooperation and connectivity. For instance, in 2015, not only did the TAR government try to position itself at the centre of the 'Bangladesh-China-India-Myanmar Economic Cooperation Belt', but also introduced the idea of the 'Himalayan Economic Cooperation Belt' (21st Century Business Herald, 2015). The latter was designed by the TAR to accelerate the opening up of the Tibetan region along the border with South Asian countries, with the focus on a cross-border economic cooperation zone, the border trade, and economic cooperation in other sectors. Three years later, the first Trans-Himalaya Economic Cooperation Forum was held in Tibet, and, in 2023, the TAR and China's Ministry of Foreign Affairs jointly organised the third 'Himalayan Rim' International Cooperation Forum with participants from over 40 countries. Meanwhile, TAR leaders have also travelled abroad to expand its regional and international engagement, though far less frequently than Yunnan's provincial leaders. In November 2023, for example, Yan Jinhai, Deputy Secretary of the Party Committee and Chairman of the TAR, led a delegation to visit the Maldives, Thailand, and Myanmar. Moreover, under the LMC, the TAR has sought opportunities to engage different stakeholders from other Mekong River countries. For instance, in October 2018, the Propaganda Departments of the TAR, Yunnan, and Qinghai jointly organised a Lancang-Mekong youth visit to various cities in China, including Chamdo City in Tibet.

Additionally, the TAR is becoming an increasingly important actor in the Mekong hydropolitics due to the geographical shift of China's dam-building activities to the upper Lancang. Now that HydroLancang has completed almost all of the dams that it had planned for the Lancang mainstream in Yunnan, it has now shifted its focus to develop hydropower on the mainstream of the Tibetan section. In April 2021, HydroLancang announced carrying out the preliminary work on the first of these (The Paper 2023). Eight dams are planned for TAR Lancang reaches (Table 2.1).

Table 2.1: Planned dams on the Tibet section of the Lancang River (Chinapower, 2022).

	Dam Name	Installed Capacity (Megawatts)
1	Sidege	129
2	Yue Long	129
3	Kagong	240
4	Banda	1,500
5	Rumei	2,600
6	Bangduo	720
7	Guxue	2,100
8	Gushui	2,200

In 2022, HydroLancang announced that the company planned to develop a 20GW hydropower and photovoltaic project on the upper Lancang River in Tibet to establish the Tibet Lancang River Clean Energy Base. A ±800kV ultra-high voltage direct current transmission line is currently under construction to deliver Tibetan power to the Guangdong, Hong Kong, and Macao Greater Bay Area (Chinapower, 2022). The construction is set to start gradually during the 14th Five-Year Plan period, and HydroLancang plans to start power transmission in 2030 and complete the project in 2035. In April 2023, HydroLancang announced that it had received approval from the National Development and Reform Commission to commence construction of a 2,600 MW hydropower plant (in Rumei) on the Lancang in the TAR (Hayley, 2023). The project is one of the three watershed water–wind–solar integration demonstration bases to be promoted by the National Energy Administration, with a total installed capacity of about 50,000 MW, and is scheduled for completion in 2032 (Chinapower, 2023). This national-level focus effectively elevates the strategic importance of Tibet, making the TAR government an increasingly important stakeholder in shaping China's hydrological and energy policies.

It is not just the HydroLancang that is interested in exploiting Tibet's hydropower resources. By about 2020, most of China's planned hydropower projects had been developed, except for those in Tibet—due to the topographic and altitudinal factors mentioned earlier. With its large water resources and also an important strategic reserve (in glaciers), Tibet has huge hydro-development potential. By the end of 2017, Tibet had an installed hydropower capacity of 1,770 MW—just 1 per cent of its technically exploitable potential (Tan, 2019). All of China's major hydropower companies have been racing to establish and expand their presence in Tibet, which has provided the TAR with more economic and political opportunities as these big state-owned hydropower companies are not only powerful political actors on their own, but also have large economic and academic resources (for instance, the engineers and researchers affiliated with Huaneng Group are the top contributors to academic articles on the Lancang-Mekong River). Over the past decades, strategic partnerships have been formed between the TAR and different hydropower companies. In 2010, for example, the TAR government signed a strategic cooperation agreement with China's Huadian Group, another major Chinese hydropower company, to jointly exploit the hydropower resources of Tibet (TAR Government, 2010).

The TAR regional government has concurrently made the development of green energies—in which it includes hydropower—as the primary driver of its economic development. Tibet's economic development has long relied on external support, with limited internal capacity for economic growth. Developing hydropower in Tibet is identified as the pivotal strategy for transforming Tibet's resource advantage into an economic one. The development of hydropower can effectively convert its resource advantage into economic and industrial prowess (Zhang and Donnellon-May, 2021a). In addition, the development of Tibet's hydropower is considered a critical pathway to significantly strengthen its socioeconomic development, thereby simultaneously supporting major national policies such as 'ecological civilisation' and 'rural revitalisation'. In the early 2010s, Tibet formulated a three-step plan to develop its electricity production and distribution sectors, which included the rapid development of its hydropower (Chinanews, 2013). This approach has proved successful for the TAR. Power consumption has grown from 2.8 million megawatt hours (MWh) in 2012 to 10.1 million MWh in 2021. At present, Tibet's main power grid covers all counties (districts) and major townships in the territory, making highly reliable power available to its population of 3.45 million people. The TAR started exporting power to other Chinese provinces in 2015, and, in 2021, dispatched 9.1 million MWh electricity (Chinanews, 2022).

Ultimately, Tibet seeks to become a new driver of economic development for Southwestern China by taking advantage of its hydropower resources, and attracting economic benefits to the territory (Wang, 2016). These initiatives find a ready ally in Chinese dam-building corporations, creating a symbiotic relationship that amplifies Tibet's relevance in Mekong hydropolitics. Over the last few decades, the Tibetan government has signed various cooperation agreements with the country's main power SOEs, to promote the development of hydropower. These partnerships serve to amplify the TAR's institutional capacity, enabling it to effectively advocate for its interests in the complex, multi-stakeholder landscape of Mekong governance.

Finally, there are strategic opportunities associated with the Chinese administration's ambitious decarbonisation goals. Under the Xi administration, China has committed to ambitious climate targets that necessitate transitioning from fossil fuels towards clean energy options, of which hydropower is perceived as one. These ambitions commit China to carbon emissions peaking by 2030 and achieving carbon neutrality before 2060. The reduction in carbon emissions and transition to clean energy stand out as a central policy priority in the 14th Five-Year Plan (BBC, 2020) which sets an 18 per cent reduction target for CO_2 intensity and a 13.5 per cent reduction target for energy intensity from 2021 to 2025. The framing of hydropower as clean energy means that, as China shifts away from coal (which supplies nearly 70 per cent of its current energy use), more dam construction can be expected.

Given that Tibet is among the last regions in China with significant unexploited hydro potential, it is becoming a focus of these national policy goals. The TAR government can leverage this to build coalitions and advance its hydropower ambitions, thus gaining more say in Mekong River hydropolitics. The TAR government's strategic vision of becoming the 'clean energy base' for China by exporting surplus electricity aims to generate revenue streams that can undergird broader economic development goals (Xinhuanet, 2022). As a key mechanism for the promotion of renewable energy production and consumption, Tibetan water resources and development are also seen to play an important role in safeguarding China's national energy supplies and security. Ever since developing hydropower, other clean energy sectors have constantly been stressed and prioritised by successive TAR governments and in numerous government policy documents. Most recently, in 2021, the TAR Government issued its 14th Five-Year Plan and a 2035 vision that calls for the accelerated development of hydropower and associated transmission infrastructure on the upper reaches of the Lancang River (TAR Government, 2021).

In summary, the geographical shift in dam construction, the national focus on hydropower development, the TAR's economic imperatives, and alignment with national climate goals are cumulatively enhancing Tibet's stake in the Mekong River. Moreover, the strategic partnerships forged by the TAR government with the major hydropower SOEs provide it with the requisite capacity to navigate the complexities of transboundary hydropolitics. As China continues to be a dominant player in this arena, the evolving role of the TAR warrants closer scrutiny for a comprehensive understanding of the future trajectory of this geopolitically significant waterway.

Discussion

While China has long been criticised for its hegemonic position in multiple international river basins that unilaterally builds dams upstream with little regard for the downstream countries (Chellaney, 2017; Han, 2017), there have been notable changes in China's Mekong River policy over the past few years. These changes are not only technical adjustments but represent a major shift in the conduct of China's Mekong River hydro diplomacy. There are multiple reasons for this (Zhang and Li, 2020). At the national level, China's high-profile BRI and the introduction of the new neighbourhood diplomacy[4] are among the key factors that have contributed to the emergence of China's variant of water diplomacy in the Mekong River Basin. Focusing on subnational actors, this chapter has described how the YPG has played a significant role in shaping China's overall foreign policy with its southern neighbours in general and the country's Mekong River policy in particular via active lobbying and close collaboration with its Mekong neighbours.

After Xi Jinping came to power in 2012, Beijing's re-centralisation push in both security and foreign policymaking weakened Yunnan's policy innovation dealing

[4] The new neighbourhood diplomacy was first proposed by Xi Jinping in 2013 and it outlines China's commitment to pursue friendship and partnership with its neighbouring countries, and to build an amicable, secure and prosperous neighbourhood.

with Mainland Southeast Asia (Lee, 2017). Nevertheless, Yunnan continues to be an important influence on China's relations with its southern neighbours. As discussed, Xi's high-profile BRI and the LMC have provided excellent opportunities for Yunnan to become the 'pivot of Southeast Asia' through greater connectivity and regional cooperation, including engagement in regional hydropower transboundary water governance. Furthermore, the YPG and its affiliated actors have reached out to national and subnational governments of the LMB countries to push forward economic relations and infrastructure development, which subsequently influence China's Mekong River policy.

In addition to the YPG, this chapter also sheds light on the emerging importance of Tibet in Mekong hydropolitics. As China commences dam construction activities on the upper Lancang River in the TAR, there are emergent regional and national hydropolitical implications. Firstly, dam construction on the Tibetan reaches of the Lancang suggests that the region is likely to exert increased influence within national and regional hydropolitical arenas. The dams in Tibet will augment its stake in wider Lancang-Mekong hydropolitics. The impacts of Tibet's dams may, however, be largely absorbed by downstream Yunnan Province. This geographical buffering could minimise international controversies surrounding Tibet's dam projects, unlike the dams in Yunnan, which have been subjected to considerable scrutiny and debate.

The initiation of dam projects not only bolsters Tibet's capacity for internal hydropower production but could also amplify its voice in trans-provincial and eventually transnational water management. This enhanced relevance is particularly salient considering longstanding proposals to divert water from transboundary rivers on the Qinghai-Tibetan Plateau to more arid regions in northern China. Notable proposals include the Shuotian Canal and the Red Flag River project, which envisage rerouting water from vital rivers like the Brahmaputra, Mekong, and Salween.

The construction of dams in Tibet is likely to shift the region's stance on such diversion projects. These dams will contribute to Tibet's economic strategy, and any diversion of water would undermine the region's hydropower potential, making it economically disadvantageous for the region. Therefore, it can be anticipated that Tibet will increasingly oppose any such plans, much like Sichuan Province and Hubei have done in the case of the western route of the South–North Water Diversion project (Zhang and Donnellon-May, 2021b). Those regions have staunchly resisted water diversion from the upper Yangtze River due to its detrimental impact on their hydropower operations.

In short, the resumption of dam construction in Tibet has the potential to bring about a realignment in its role within the broader Lancang-Mekong hydropolitics—particularly if its hydropower construction is perceived to yield downstream impacts amongst the lower Mekong countries. Tibet is likely to evolve from a peripheral player into a more active stakeholder, affecting both national and regional hydropolitics. The reinitiation of dam projects will not only enhance Tibet's influence but also align it more closely with economic imperatives that are divergent from longstanding water diversion proposals. The nuanced interplay of these factors necessitates a re-evaluation of Tibet's emerging role and its implications for the broader hydropolitical landscape of the MRB.

Conclusion

These findings show that—even though China is a unitary government, in the view of many, it has an authoritarian regime after years of recentralisation—Chinese subnational governments, especially provincial governments, should not be understood as mere extensions of the central government that are unable to act according to their priorities in the context of transboundary hydropolitics. Rather, Chinese provincial governments have played important roles not only in implementing policies and rules set by national bodies but also, more importantly, in shaping Chinese engagement with Indochina around water resources development.

References

21st Century Business Herald. 2015. Dǎzào huán xǐmǎlāyǎ jīngjì hézuò dài xīzàng yíngxīn zhèngcè chuāngkǒu [Building a Himalayan Economic Cooperation Ring and Tibet's New Policy Window]. *21st Century Business Herald* 4 June 2016, https://m.21jingji.com/article/20150604/79bbcfddc764257fe99fbd9ed25e62bb.html (accessed 29 September 2023).

BBC. 2020. Climate change: China aims for 'carbon neutrality by 2060.' *BBC News* 22 September 2020, https://www.bbc.com/news/science-environment-54256826 (accessed 24 September 2023).

Biba, S. 2012. China's continuous dam-building on the Mekong River. *Journal of Contemporary Asia* 42(4): 603–628, https://doi.org/10.1080/00472336.2012.712257.

Biba, S. 2016. From securitization moves to positive outcomes: the case of the spring 2010 Mekong crisis. *Security Dialogue* 47(5): 420–439, https://doi.org/10.1177/0967010616657795.

Biba, S. 2018. China's 'old' and 'new' Mekong River politics: the Lancang-Mekong Cooperation from a comparative benefit-sharing perspective. *Water international* 43(5): 622–641, https://doi.org/10.1080/02508060.2018.1474610.

Brauer, R. and Kliem, F. 2017. Coercive water-diplomacy: playing politics with the Mekong. International Reports 3. Berlin: Konrad-Adenauer-Stiftung, https://www.kas.de/documents/252038/253252/7_dokument_dok_pdf_50363_2.pdf/c82d2f4e-6ef5-3bad-c444-38e7739f7637?version=1.0&t=1539655675479 (accessed 28 March 2024).

Busbarat, P., Bunyavejchewin, P. and Suporn, T. 2021. China and Mekong regionalism: a reappraisal of the formation of Lancang-Mekong Cooperation. *Asian Politics & Policy* 13(2): 193–211, https://doi.org/10.1111/aspp.12575.

Chen, H., Rieu-Clarke, A. and Wouters, P. 2013. Exploring China's transboundary water treaty practice through the prism of the UN Watercourses Convention. *Water International* 38(2): 217–230, https://doi.org/10.1080/02508060.2013.782134.

Chellaney, B. 2017. A new front in Asia's Water war. Project Syndicate, https://www.project-syndicate.org/commentary/china-no-hydrological-data-for-india-by-brahma-chellaney-2017-10 (accessed 28 May 2024).

Chinanews. 2009. Zhōngguó shuǐ néng kāifā zhòngdiǎn xī yí zàngqīng diān huò chéng zhòngdiǎn [China's hydropower development focus moves westward, Tibet, Qinghai and Yunnan may become the focus]. *Chinanews* 11 August 2009, https://www.chinanews.com.cn/cj/cj-gncj/news/2009/08-11/1812047.shtml (accessed 24 September 2023).

Chinanews. 2013. Qīngzàng liánwǎng gōngchéng mǎn zhōusuì diànlì tiān lù liàng xuě yù-zhōng xīn wǎng. [The Qinghai-Tibet Power Grid Project Celebrates Its First Anniversary: The Electric Highway Brightens the Snowy Plateau.] *Chinanews* 4 January 2013, https://www.chinanews.com/ny/2013/01-04/4454962.shtml (accessed 24 September 2023).

Chinanews. 2022. Zhōngguó zhè shí nián) xīzàng diànwǎng shí nián lěijì wánchéng jiànshè tóuzī jìn 700 yì yuán [In the past ten years (China), the Tibetan power grid has completed construction investment of nearly 70 billion yuan in ten years]. *Chinanews* 4 October 2022, https://m.chinanews.com/wap/detail/chs/zw/9866631.shtml (accessed 24 September 2023).

Chinapower. 2022. Huánéng shuǐdiàn: Nǐ tóuzī 50 yì dǎzào "shuāng qiān wàn qiānwǎ" qīngjié néngyuán jīdì!_Diànlìwǎng [Huaneng Hydropower: Plans to invest 5 billion to build a 'double 10 million kilowatt' clean energy base!]. *Chinapower* 15 March 2022, http://mm.chinapower.com.cn/tynfd/hyyw/20220315/138765.html (accessed 24 September 2023).

Chinapower. 2023. Láncāngjiāng xīzàng duàn shuǐ fēngguāng yītǐ huà qīngjié néngyuán jīdì jì RM shuǐdiàn gōngchéng zhèngshì kāigōng [Construction of the water, wind and solar integrated clean energy base and RM hydropower project in the Tibet section of the Lancang River officially started]. *Chinapower* 29 May 2023, http://mm.chinapower.com.cn/guihuajianshe/qiye/2023-05-29/202355.html (accessed 24 September 2023).

Gerlak, A.K. and Haefner, A. 2017. Riparianization of the Mekong River Commission. *Water International* 42(7): 893–902, https://doi.org/10.1080/02508060.2017.1376267.

Goh, E. 2004. China in the Mekong River Basin. Working Paper 69. Singapore: Institute of Defence and Strategic Studies Singapore, https://www.rsis.edu.sg/wp-content/uploads/rsis-pubs/WP69.pdf (accessed 28 March 2024).

Grumbine, R.E., Dore, J. and Xu, J. 2012. Mekong hydropower: drivers of change and governance challenges. *Frontiers in Ecology and the Environment* 10(2): 91–98, https://doi.org/10.1890/110146.

Guo, Y. 2017. The evolution of China's water diplomacy in the Lancang-Mekong River Basin: motivation and policy choices. In Zhang, H. and Li, M. (eds.), *China and Transboundary Water Politics in Asia*, 82–98. Abingdon: Routledge.

Han, H. 2017. China, an upstream hegemon: a destabilizer for the governance of the Mekong River? *Pacific Focus* 32(1): 30–55, https://doi.org/10.1111/pafo.12085.

Hayley, A. 2023. China's Huaneng Group plans 2.6GW hydro project in Tibet. *Reuters* 30 May 2023, https://www.reuters.com/business/energy/chinas-huaneng-group-plans-26gw-hydro-project-tibet-2023-05-30/ (accessed 24 September 2023).

Hecht, J.S., Lacombe, G., Arias, M.E., Dang, T.D. and Piman, T. 2019. Hydropower dams of the Mekong River basin: a review of their hydrological impacts. *Journal of Hydrology* 568(2019): 285–300, https://doi.org/10.1016/j.jhydrol.2018.10.045.

Ho, S. 2014. River politics: China's policies in the Mekong and the Brahmaputra in comparative perspective. *Journal of Contemporary China* 23(85): 1–20, https://doi.org/10.1080/10670564.2013.809974.

Ho, S. 2017. Introduction to 'Transboundary River Cooperation: Actors, Strategies and Impact.' *Water International* 42(2): 97–104, https://doi.org/10.1080/02508060.2017.1279042.

Kondolf, G.M., Schmitt, R.J.P., Carling, P., Darby, S., Arias, M., Bizzi, S., Castelletti, A., Cochrane, T.A., Gibson, S., Kummu, M., Oeurng, C., Rubin, Z. and Wild, T. 2018. Changing sediment budget of the Mekong: cumulative threats and management strategies for a large river basin. *Science of The Total Environment* 625(2018): 114–134, https://doi.org/10.1016/j.scitotenv.2017.11.361.

Lee, S. 2017. An institutional analysis of Xi Jinping's centralization of power. *Journal of Contemporary China* 26(105): 325–336, https://doi.org/10.1080/10670564.2016.1245505.

Li, M. 2014. Local liberalism: China's provincial approaches to relations with Southeast Asia. *Journal of Contemporary China* 23(86): 275–293, https://doi.org/10.1080/10670564.2013.832530.

Liu, T. and Song, Y. 2019. Beyond the hinterland: exploring the international actorness of China's Yunnan province. *International Relations of the Asia-Pacific* 21(3): 335–370, https://doi.org/10.1093/irap/lcz026.

McDonald, K., Bosshard, P. and Brewer, N. 2009. Exporting dams: China's hydropower industry goes global. *Journal of Environmental Management* 90: S294–S302, https://doi.10.1016/j.jenvman.2008.07.023.

MERFI (Mekong Region Futures Institute). 2025. Dataset on the dams of the Greater Mekong. Bangkok: MERFI, https://www.merfi.org/mekong-region-dams-database (accessed 25 April 2025).

Middleton, C. and Devlaeminck, D.J. 2021. Reciprocity in practice: the hydropolitics of equitable and reasonable utilization in the Lancang-Mekong basin. *International Environmental Agreements: Politics, Law and Economics* 21(2): 235–253, https://doi.org/10.1007/s10784-020-09511-6.

MRC (Mekong River Commission) and MoWRoC (Ministry of Water Resources of China). 2016. Joint observation and evaluation of the emergency water supplement from China to Mekong River. Vientiane: Mekong River Commission, https://pdfs.semanticscholar.org/f2a8/dcf656bbce4f-08c71645d26d7ea9a5f4fe28.pdf (accessed 13 September 2024).

Pohl, B., Carius, A., Conca, K., Dabelko, G.D., Kramer, A., Michel, D., Schmeier, S., Swain, A. and Wolf, A. 2014. The rise of hydro-diplomacy: strengthening foreign policy for transboundary waters. Berlin: ADELPHI, https://www.adelphi.de/en/system/files/mediathek/bilder/the_rise_of_hydro-diplomacy_adelphi.pdf (accessed 20 July 2018).

Qiao, X. 1996. Zhuazhujiyu jiakuaibufa tuidong LancangJiang MeigongHe guojihangyun kaifa [Seize the opportunity and speed up the pace to promote the development of international shipping on the Lancang-Mekong River]. *Pearl River Water Transport* 12(1996): 5–7.

Ren, J. and Li, L. 2018. *Moshao bian qianyan Yunnan mianxiang xinan kaifang jishi* [The last frontier: a chronicle of Yunnan's opening to the Southwest]. 1st ed. Kunming: Yunnan Daily Press Group.

SCMP. 2024. China says Mekong dam did not discharge water downstream amid heavy flooding in Thailand. *South China Morning Post* 28 August 2024, https://www.scmp.com/news/china/diplomacy/article/3276291/china-says-mekong-dam-did-not-discharge-water-downstream-amid-heavy-flooding-thailand (accessed 13 September 2024).

Spijkers, O. and Devlaeminck, D.J. 2022. Layers of regulation in transboundary water governance: exploring the role of third states in the Lancang–Mekong. *Water International* 47(1): 132–151, https://doi.org/10.1080/02508060.2021.1960100.

Tan, J. 2019. Cùjìn xuě yù gāoyuán jīngjì shèhuì huánjìng xiétiáo fāzhǎn tuījìn néngyuán lǜsè gémìng, bǎozhàng guójiā néngyuán ānquán——xīzàng shuǐdiàn fāzhǎn lìchéng hé fāzhǎn zhǎnwàng. Shuǐlì fādiàn 2. [Promote the coordinated development of the economy, society and environment in the snow-covered plateau, promote the green energy revolution, and ensure national energy security – the development history and prospects of hydropower in Tibet]. *Hydropower* 45(2): 29–34, https://doi.org/10.3969/j.issn.0559-9342.2019.02.009.

TAR Government. 2010, Xīzàng zìzhìqū zhèngfǔ yǔ zhōngguó huádiàn jítuán gōngsī jǔxíng zuòtán_xīzàng zìzhìqū rénmín zhèngfǔ [Tibet Autonomous Region Government held a discussion with China Huadian Corporation]. 10 November 2010, https://www.xizang.gov.cn/xwzx_406/bmkx/201812/t20181216_17602.html (accessed 24 September 2023).

TAR Government. 2021. Zhōnggòng xīzàng zìzhìqū wěiyuánhuì guānyú zhìdìng guómín jīngjì hé shèhuì fāzhǎn "shísìwǔ" guīhuà hé èr líng sānwǔ nián yuǎnjǐng mùbiāo dì jiànyì [Recommendations of the Tibet Autonomous Region Committee of the Communist Party of China on formulating the '14th Five-Year Plan' for national economic and social development and long-term goals for 2035], https://www.xizang.gov.cn/zwgk/xxfb/gsgg_428/202206/W020220623689652560071.pdf (accessed 24 December 2023).

The Laotian Times. 2022. Nam Tha 1 Hydropower Station successfully sends power to China. *The Laotian Times* 22 June 2022, https://laotiantimes.com/2022/06/22/nam-tha-1-hydropower-station-successfully-sends-power-to-china/ (accessed 13 July 2024).

The Paper. 2023. Huánéng shuǐdiàn: Nǐ zài xīzàng dìqū tóuzī jiànshè shuǐdiànzhàn, zǒng tóuzī chāo 580 yì yuán [Huaneng Hydropower: Plans to invest in the construction of hydropower stations in Tibet, with a total investment of over 58 billion yuan]. *The Paper* 30 May 2023, https://m.thepaper.cn/newsDetail_forward_23286177 (accessed 24 December 2023).

Tian, F. and Liu, H. 2016. 'China's water hegemony in Asia' or 'shared rivers, shared future': which? Embassy of the People's Republic of China in the Kingdom of Thailand, 27 July 2016, http://th.china-embassy.gov.cn/eng/ztbd/201607/t20160727_10148408.htm (accessed 13 September 2024).

Tubilewicz, C. 2017. Paradiplomacy as a provincial state-building project: the case of Yunnan's relations with the Greater Mekong Subregion. *Foreign Policy Analysis* 13(4): 931–949, https://doi.org/10.1093/fpa/orw022.

Wang, Q. 2016. Guānyú xīzàng pínkùn dìqū zài shuǐ néng kāifā zhōng kāizhǎn shuǐdiàn zīchǎn shōuyì fúpín de sīkǎo [Reflections on Using Hydropower Asset Benefits for Poverty Alleviation in Impoverished Areas of Tibet]. *Tibet Technology* 2016 (10): 34–35, https://doi/10.3969/j.issn.1004-3403.2016.10.012.

Wang, R.Y., Liu, X., Zhang, W. 2022. China's water governmentality and the shaping of hydrosocial territories in the Lancang-Mekong Region. *The China Quarterly* 252(2022): 1233–1255, https://doi.org/10.1017/S0305741022000820.

Wong, A. 2018. More than peripheral: how provinces influence China's foreign policy. *China Quarterly* 235(2018): 735–757, https://doi.org/10.1017/S0305741018000930.

Xinhuanet. 2022. Tibet powers up clean energy development. *Xinhuanet* 30 December 2022, https://english.news.cn/20221230/3405b61320834248bae871f27b156bac/c.html (accessed 24 December 2023).

Yeophantong, P. 2017. River activism, policy entrepreneurship and transboundary water disputes in Asia. *Water International* 42(2): 163–186, https://doi.org/10.1080/02508060.2017.1279041.

YPTD (Yunnan Provincial Transportation Department). 2022. Yúnnán shěng shuǐlù jiāotōng "shísìwǔ" fāzhǎn guīhuà [Yunnan Province Water Transport "14th Five-Year Plan" Development Plan], https://www.yn.gov.cn/ztgg/ynghgkzl/sjqtgh/zxgh/202202/P020230829849312784104.pdf (accessed 23 May 2023).

Zhang, H. 2016. Sino-Indian water disputes: the coming water wars? *WIRE's Water* 3(2): 155–166, https://doi.org/10.1002/wat2.1123.

Yuwei, H. 2022. Envoys, experts vow further cooperation amid challenges in the Lancang-Mekong region. *Global Times* 31 May 2022, https://www.globaltimes.cn/page/202205/1267090.shtml (accessed 13 September 2024).

Zhang, H. and Donnellon-May, G. 2021a. China's hydropower plan on the Brahmaputra. *The Diplomat* 1 September 2021, https://thediplomat.com/2021/09/chinas-hydropower-plan-on-the-brahmaputra/ (accessed 17 February 2022).

Zhang, H. and Donnellon-May, G. 2021b. To build or not to build: Western route of China's South-North Water Diversion Project. *New Security Beat* 12 August 2021, https://www.newsecuritybeat.org/2021/08/build-build-western-route-chinas-south-north-water-diversion-project/ (accessed 17 February 2022).

Zhang, H. and Li, M. (eds.). 2017. *China and transboundary water politics in Asia.* 1 edition. New York (NY): Routledge.

Zhang, H. and Li, M. 2020. China's water diplomacy in the Mekong: a paradigm shift and the role of Yunnan provincial government. *Water International* 45(4): 347–364, https://doi.org/10.1080/02508060.2020.1762369.

Zhang, L. and Lu, G. 2016. Transboundary water cooperation from the perspective of water release under LMC. *World Outlook* (5): 95–112 (in Chinese).

Zhang, L. and Zhang, H. 2021. Water diplomacy and China's bid for soft power in the Mekong. *The China Review* 21(4), 39-75, https://cup.cuhk.edu.hk/image/catalog/journal/jpreview/CR21.4_39-75.pdf (accessed 29 March 2026).

Zhong, Y., Tian, F., Hu, H., Grey, D. and Gilmont, M. 2016. Rivers and reciprocity: perceptions and policy on international watercourses. *Water Policy* 18(4): 803–825, https://doi.org/10.2166/wp.2016.229.

CHAPTER 3

Situating the mandate of the Mekong River Commission within a changing hydrosocial territory and regional order

Ming Li Yong[1]

Abstract

There has been much debate about the mandate of the Mekong River Commission (MRC), generally rooted in a multi-scalar tension between managing state-led water resources development projects in a way that ensures the equitable and reasonable use of water between Mekong nations as well as safeguarding the ecological integrity of the river. This chapter traces the shifting hydrosocial territory of the Mekong River back to diverging interpretations of the MRC's mandate, the limitations of this mandate, and how it has facilitated the creation of a powershed in the Lower Mekong Basin. The chapter considers the MRC's ability to maintain its relevance by navigating development challenges, climate change, and geopolitical contestations amidst drastic biophysical changes.

Keywords: Mekong River Commission, hydroterritories, transboundary water governance, Mekong River

Introduction

On 4 April 2023, the Chief Executive Officer (CEO) of the Mekong River Commission (MRC) Secretariat, Dr Anoulak Kittikhoun, gave his second State of the Mekong Address since stepping into the position in 2022. In his opening remarks, he said:

> In two days it is Mekong Day. On that day last year, I gave the inaugural Mekong address. The idea was to tell as it is, what's happening in the Mekong, what's being done about it, given the mandate and the role of the MRC. I say that very carefully, given the mandate and the role of the MRC, as we cannot do everything. The hope is to inspire you in your respective roles and interests to complement our interests, to fill our gaps, to do what we cannot (2023).

[1] East-West Center, Honolulu, Hawaii, USA.

The CEO's careful emphasis on the MRC's limited mandate can be situated within the contentious debates that have occurred over the MRC's ability to manage the transboundary Mekong River for the benefit of its inhabitants. The MRC is an intergovernmental organisation comprising Laos, Thailand, Cambodia, and Vietnam, whose mandate is articulated in the *Agreement on the Cooperation for the Sustainable Development of the Mekong River Basin* (henceforth, the '1995 Mekong Agreement'). Article 1 of the 1995 Agreement states that the parties will "cooperate in all fields of sustainable development, utilization, management and conservation of water and related resources of the Mekong River Basin" (MRC, 1995: 6). The Mekong River Basin's total estimated population of ~72 million is still predominantly rural (MRC, 2021a), and the Mekong River's natural resources provide a critical foundation for their food security and livelihoods. However, hydropower development on China's section of the river (also known as the Lancang River) and the accelerated pace of hydropower development in the Lower Mekong Basin (comprising Laos, Thailand, Cambodia, and Vietnam) in the past two decades has contributed towards a perceived ineffectiveness of the MRC to resolve disputes between stakeholders, and concerns over the potentially devastating impacts on the food security and livelihoods of the millions of people dependent on the Mekong River's ecosystem services.

While the roles and responsibilities of the MRC extend across a range of functions, this chapter focuses on the organisation's mandate and procedures in relation to hydropower development on the Mekong River's mainstream. It is important to note that the MRC's mandate, while broadly set out in the 1995 Agreement, is not static. Rather, varying stakeholder interpretations and ambiguity surrounding the Agreement have given rise to diverse assertions on the role that the MRC should play in transboundary water governance. These contestations can be situated within a recognition that the MRC is not a homogeneous entity (Dore and Lazarus, 2009). In addition, a range of stakeholders from within and beyond the MRC have sought in varied ways to influence the governance of the Mekong River basin.

This chapter situates the MRC's contested and limited mandate in three periods that reflect shifts in the regional order and forms of 'regionness' in the Mekong River Basin. First, the chapter discusses the formation of the Mekong Committee and geopolitical and geoeconomic shifts leading up to the formation of the MRC and the weakening of its mandate. Second, the limitations of the MRC's mandate are demonstrated through a focus on the Procedures for Notification, Prior Consultation and Agreement (PNPCA), which have facilitated the creation of a 'powershed' and fragmented watershed in the Lower Mekong Basin. Third, the role of China and the Lancang Mekong Cooperation (LMC) mechanism is considered within a changing regional order, and the implications for the MRC will be discussed in the context of geostrategic competition and the drastic biophysical changes occurring to the river.

Through the perspectives of hydrosocial territories and critical institutionalism, this chapter seeks to understand how the shifting and contested mandate of the MRC is tied up with constructions of the 'region'.

The governance arrangements of the MRC

The MRC describes itself as "an intergovernmental body working to facilitate cooperation on the sustainable development and management of the Mekong River Basin", acting as a "platform for water diplomacy and regional cooperation" even where national interests diverge, and as a "knowledge hub that promotes evidence-based policy making" (MRC, n.d.(a): 1). The MRC's initiatives cover a range of thematic areas including agriculture, basin development planning, flood and drought management, fisheries, and sustainable hydropower (MRC, n.d.(b)). The MRC comprises three permanent bodies: The Council, Joint Committee (JC), and the MRC Secretariat. The Council is the top decision-making body in the MRC and comprises one member from each riparian state at the ministerial and cabinet level, who is "empowered to make policy decisions on behalf of his/her government" (MRC, 1995: 6). At the time of writing in January 2024, the Council comprises ministers from the Ministries of Natural Resources and Environment (MONRE) in Laos and Vietnam, a deputy prime minister who chairs the Thai National Mekong Committee, and the Ministry of Water Resources and Meteorology (MOWRAM) in Cambodia. These ministries tend to be less powerful compared to the key energy, finance, and planning ministries in their countries that play a role in planning and approving key projects on the Mekong River and its tributaries. The ability of MONRE or MOWRAM ministers to make policy decisions that diverge from the broader national interests of their governments is, therefore, limited (Boer et al., 2015).

The second body making up the MRC is the JC, which implements initiatives from the Council and also supervises the activities of the MRC Secretariat (Lee and Scurrah, 2009). The JC comprises one member from each riparian state with a rank of departmental head or higher. The third body in the MRC is the MRC Secretariat, which is the technical and administrative arm of the MRC. The MRC Secretariat renders assistance to the Council and JC and the carrying out of its day-to-day work (Boer et al., 2015), including the implementation of MRC procedures such as the PNPCA. The MRC Secretariat is headed by a Chief Executive Officer (CEO), who is appointed by the Council from a shortlist of candidates selected by the JC. One of the main functions of the MRC Secretariat has been to run key programmes listed earlier. The MRC also works closely with actors who are not specifically mentioned in the 1995 Mekong Agreement. Each member state has a National Mekong Committee (NMC), which serves as a coordinating agency between the MRC and the relevant government line agencies in their respective countries. These agencies may include those relating to water, fisheries, energy, planning, and foreign affairs. The NMCs are served by a Secretariat, who, like the MRC Secretariat, is the administrative and technical arm that carries out day-to-day work alongside duties relating to "cross-sectoral, cross-agency coordination, communication and reporting" (MRC, n.d.(c)).

The MRC also coordinates with a Donor Consultative Group, comprising agencies which administer official development assistance from various governments[2] and the International Union for Conservation of Nature (IUCN). These 'development partners' (DPs) have historically provided funding to the Secretariat. In 2009, donor funding comprised 90 per cent of the MRC's funding (Lee and Scurrah, 2009). The MRC has since set a goal to be fully self-financed by its member countries by 2030 and to move towards "full riparianisation[3] of the MRC and to the decentralisation of the organisation's core functions to Member Countries… and an associated shift in the MRC Secretariat to a leaner organisation focused on the MRC's core functions" (MRC, 2019a: 169). This direction was set in 2010 at the inaugural MRC Summit between the Prime Ministers of its member countries in Hua Hin. However, in 2016, the MRC saw funding from DPs slashed by more than half, from USD 115 million for 2011-2015 to just USD 53 million for 2016-2020, which was perceived by some as a reflection of the lack of trust that certain donors had in the MRC to effectively manage development challenges along the river (Kossov and Lay, 2016; Wright, 2016). During this time, the MRC reduced its staff number by almost 50 per cent, delegated functions such as water monitoring to member states, and downsized the MRC Secretariat from 12 programmes to four core functions (Gerlak and Haefner, 2017). Given that the MRC was often hamstrung by the national interests of its member countries, this move towards riparianisation was somewhat controversial.

Hydrosocial territories and critical institutionalism

Hydrosocial territories are defined by Boelens et al. as:

> the contested imaginary and socio-environmental materialization of a spatially bound multi-scalar network in which humans, water flows, ecological relations, hydraulic

[2] Australia, Flanders/Belgium, France, Germany, Japan, Luxembourg, New Zealand, Sweden, Switzerland, The European Union, and the United States.

[3] Initially, key technical positions within the MRC, including the CEO, were reserved for foreign staff, ostensibly for reasons of neutrality. 'Riparianisation' refers to the replacement of permanent foreign staff with professions from its member states. In 2016, the first regional CEO was appointed (Gerlak and Haefner, 2017).

> infrastructure, financial means, legal-administrative institutions and practices are interactively defined, aligned and mobilized through epistemological belief systems, political hierarchies and naturalizing discourses (2016: 2).

These territories are further constituted by hydrosocial networks comprising heterogeneous elements, which vary in their span or spatial scale, and their durability which reflects their stability over time (Boelens et al., 2016). Hydrosocial territories are also socionatural hybrids whereby "both society and nature are produced, and are hence malleable, transformable, and potentially transgressive" (Swyngedouw, 1999: 446). These territories are entangled, and "are not just bounded, material territorial projects, but also *imagined* water-based territories" (Mills-Novoa et al., 2020: 92) that are also "mediated in everyday praxis" (Boelens et al., 2016: 8).

These hydrosocial territories, in the context of a transboundary river basin, can also be understood through critical hydropolitics. Rogers et al. (2023) note that this has involved an attentiveness to the diffused nature of power between and within multiple stakeholder groups and a knowledge politics that interrogates the logics of river basin processes. This is closely tied to the critical geopolitics literature, which examines the 'incoherencies and contradictions' of state power, and "multilayered socioeconomic and cultural processes through which 'regionness' is produced and sustained" (Kuus, 2017: 6). Regions can also be understood as social constructs (Glassman, 2010) that arise through the formation of hydrosocial territories characterised by 'territorial pluralism' (Flaminio et al., 2022), and driven by a wide array of interests, worldviews, and power dynamics. This has also occurred through the scalar construction of the Mekong River basin as a naturalised economic region by powerful state interests, and has created contestations over water resources through the reshaping of power relationships within and between stakeholder groups (Sneddon and Fox, 2006). As Dore and Lebel (2010) also point out, a politics of scale and level arise in the Mekong Region as different actors prioritise and exert their power over different spatial and temporal scales in practice, and competing scales may exist in terms of administration, hydrology, ecosystems, and the economy. In addition, the Mekong River basin can be situated within "multiple constructions of the region", such as a river, a river basin, a development arena, and an arena of contested development (Hirsch, 2016: 63).

The MRC plays a role in the production of hydrosocial territories, and its mandate as established through the 1995 Mekong Agreement offers insights into how the hydrosocial territories of the Mekong River basin have been shaped over time. This perspective can be enhanced through an understanding of *critical institutionalism*, which places an emphasis on how institutions play a dynamic role in mediating relationships between people, natural resources, and society (Cleaver and de Koning, 2015). An institutional geography considers their relationships to wider cultural, political, and economic processes, along with the distinct social and spatial relationships that arise between institutions, peoples, land uses, and resources (Philo and Parr, 2000; Billo and Mountz, 2016). Cleaver and de Koning (2015: 6) suggest that the historical trajectories of institutions are important in terms of how their 'sedimented layers of governance arrangements' influence their functions today, and that change often takes place at the 'messy middle', or meso level, of institutions. The hydrosocial territories of the Mekong River basin, as mediated by the MRC, are intertwined with a politics of truth and knowledge (Rose, 1999; Boelens et al., 2016) particularly as diverse actors work to implement and navigate the MRC's mandate in response to emerging forms of hydrosocial territories. In particular, this chapter examines how the limited mandate of the MRC has played a role in this.

Constructing the Lower Mekong basin as a region: the MRC's weakened mandate

It is possible to discern the 'Mekong River basin' as a "unit for territorialised governance, overlapping with spaces of sovereignty" (Hirsch, 2016: 68), and one that has also come to be defined through various iterations of the MRC's mandate. The Mekong River basin can be seen as a natural watershed, but it has also been defined by the powerful geopolitical imaginary of riparian states bounded by the natural watershed. For example, the division of the river basin into the 'upper' basin in China and 'lower' basin is a reflection of its geomorphological characteristics, but the 'lower' basin has also come to be defined through the 'lower' riparian states' (Laos, Thailand, Cambodia, Vietnam) membership of the MRC. In this sense, this hydrosocial territory comprises overlapping administrative and natural watershed boundaries: the territorial boundaries of each nation state do not fall squarely into the natural watershed, which only comprises to varying degrees part of a country's total territory (see Table 3.1). It is possible to discern how the hydrosocial territory of the river basin has come into being through the MRC's contested mandate and the shifting geopolitical and geoeconomic imaginations associated with the river.

The predecessor of the MRC, the Mekong Committee (MC), was formed in October 1957, sponsored by the United Nations Economic Commission for Asia and the Far East (UNECAFE); China was excluded as it was not a UN member at the time and Burma had no interest in participating. The MC had the mandate under the 1957 Statute to "promote, coordinate, supervise, and control" the planning of water resources development (Browder and Ortolano, 2000: 505–506). This was characterised by much optimism that the MC would set a precedent for large-scale river basin development planning in Asia (Jacobs, 1995), driven by the UNECAFE's and the US Bureau of Reclamation's vision to "tame the Mekong River for irrigation and hydropower development" (Middleton, 2016: 210). This hydrosocial territory was created through Cold War interests to stave off the encroachment of communism and

Table 3.1: Territory within the catchment of the six Mekong River Basin countries (MRC, 2003a)

	Country or province						
	Yunnan	**Myanmar**	**Laos**	**Thailand**	**Cambodia**	**Vietnam**	**MRB**
Area (km^2)	165,000	24,000	202,000	184,000	155,000	65,000	795,000
Catchment as % of country or province	38	4	97	36	86	20	
Catchment as % of Mekong River Basin	21	3	25	23	20	8	100

framing the Mekong River as a common regional good to support economic development. The ability of member states to unilaterally exploit the river was restricted under the 1975 Joint Declaration issued by the MC in preparation for the development of mainstream dam projects, which specified that "all mainstream, major tributary and inter-basin diversions would require the unanimous approval of the Mekong Committee prior to implementation" (Browder and Ortolano, 2000: 508).

However, the 1970s saw rising political tensions and conflict in the region and put a halt to the MC's activities and ambitions (Jacobs, 1995), and gave rise to the creation of the Interim Mekong Committee (IMC) under a declaration signed by Laos, Thailand, and Vietnam in 1978. The absence of Cambodia from the IMC led to the sidelining of grand regional plans for a dam and reservoir cascade along the Mekong River. The 1978 IMC Declaration therefore "represented a new constitutional framework for the Mekong regime, one that reflected new geopolitical realities" (Browder and Ortolano, 2000: 510) with a mandate that was weaker than that held by the IMC's predecessor agency. Under the 1978 Declaration, the IMC primarily served as a vehicle to promote water resources development and seek technical and financial assistance from donor countries, rather than to control the planning of water resources development (Browder and Ortolano, 2000). The hydrosocial territory was thus discursively and materially altered, and strategically mobilised by Thailand and Vietnam to have continued access to international assistance and to further domestic water resources projects, such as Thailand's Kong-Chi-Mun project in the northeast of the country (Browder and Ortolano, 2000).

In 1992, the Asian Development Bank (ADB) launched the Greater Mekong Subregion (GMS) Economic Cooperation Programme. The GMS, territorially 'bounded' by China's Yunnan Province and Guangxi Autonomous Region, Myanmar, Thailand, Laos, Cambodia, and Vietnam, articulated former Thai Prime Minister's

Chatchai Choonhaven's 1989 vision to transform the region from 'battlefields into marketplaces'. The Mekong River basin's hydrosocial territory thus came to be dominated by this "geopolitical vision of the Mekong as an engine of regional development" (Sneddon and Fox, 2006: 187), which called for 'subregional' cooperation and connectivity across locales extending far beyond the Mekong's waterways through economic corridors. The GMS came to replace the Mekong Committee as the principal framework for channelling economic development assistance into countries in the river basin (Middleton et al., 2009), and encouraged the development of integrated regional energy markets whereby cross-border and subregional power trading could optimise the interconnections between energy supply and demand across the region. It was within this economic climate that the Mekong Agreement was signed on 5 April 1995 by the four lower Mekong countries in Chiang Rai, Thailand, bringing the MRC into existence. China and Myanmar were invited to join the MRC but declined. The 1995 Mekong Agreement represented a major shift in the MRC's mandate, given that its "water use principles were less restrictive compared to the 1975 Joint Declaration" (Mirumachi, 2011: 94) to reflect the prioritisation of national economic development over regional river basin planning and development. The negotiations leading up to the signing of the 1995 Mekong Agreement were heavily contested by Thailand, who insisted that the veto clause in the 1975 Joint Declaration be removed so that it could potentially pursue a massive water diversion project; compromises also had to be made between the national positions of the other lower Mekong countries (Browder, 2000). This weakened mandate of the 1995 Mekong Agreement now reflected how the river basin as a hydrosocial territory had become spatially and temporally segmented through the river's mainstream and tributaries, wet and dry season use, and the location of water use (Browder, 2000). The weakening of this mandate would have strong implications for the ways in which the hydrosocial territory of the Mekong River basin would be physically and discursively altered through hydropower development.

The Limits of a mandate: The PNPCA

At the time of writing, on the river's mainstream, there were 12 operational dams in China, two operational and seven planned dams in Laos, and two planned dams in Cambodia (see Fig. 3.1). There are more than 70 operational dams in Laos, with about 30 under construction and over 200 more planned in Laos (Stimson Center, 2020). This resurgence of mainstream hydropower development following prolonged political instability in the region and the 1997 Asian Financial Crisis can be situated within the shifting regional geopolitics of the region, which are in part constituted by complex and varied discursive-material constructions of the region

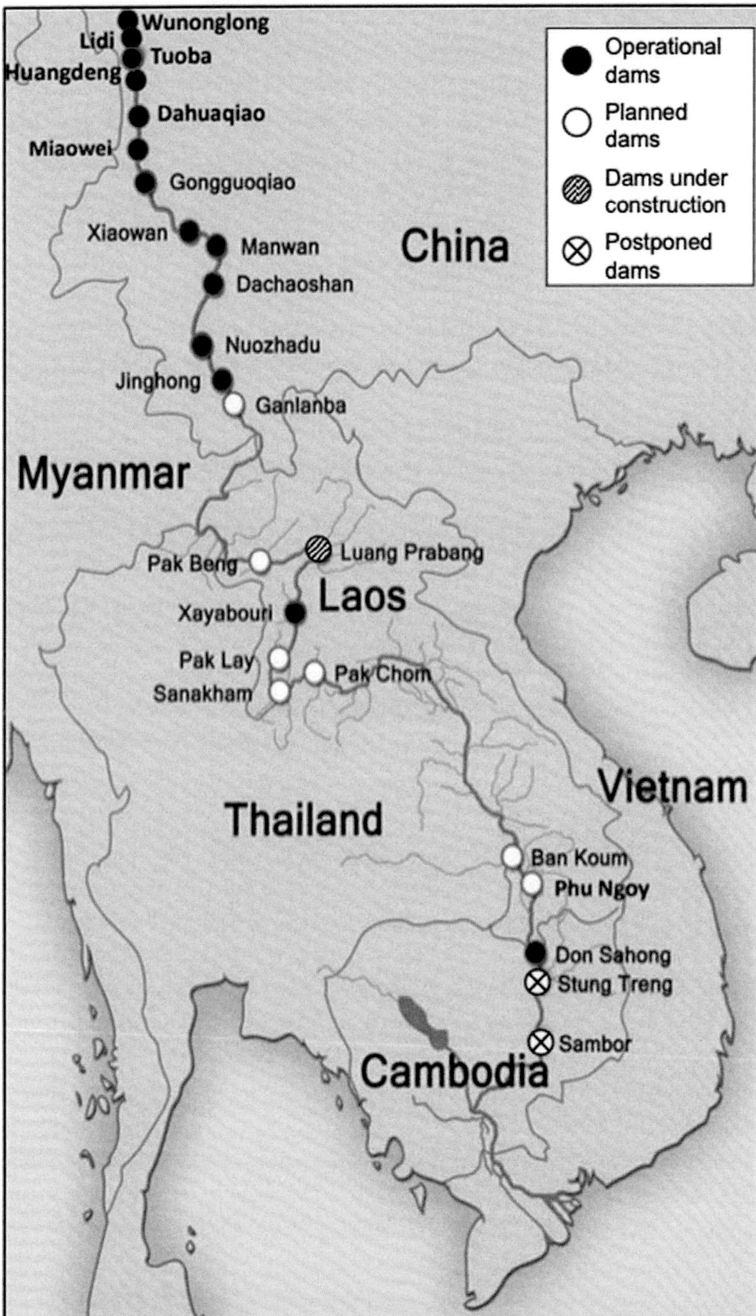

Figure 3.1: Map of hydropower dams on the Mekong River's mainstream and their status as of February 2024 (Source: adapted from Stimson Center, 2024).

(Hirsch, 2016). The hydrosocial territories of the Mekong River basin, especially in the lower basin, are mediated through the MRC's practices, particularly through the ways in which the PNPCA has been negotiated in relation to mainstream hydropower dams and the MRC's weak mandate. The perspectives of critical institutionalism shed light on the dynamics through which the MRC mediates the dangerous transformation of a transboundary ecosystem and hydrosocial territory that is increasingly physically fragmented through dam construction and operation.

The 1995 Mekong Agreement was intentionally broad so that negotiators would not get hung up over technical details (Browder, 2000). Article 5 of the 1995 Mekong Agreement emphasises the principle of 'reasonable and equitable' utilisation, which is thus supplemented by five procedures developed for data sharing, water use monitoring, maintenance of flows on the river's mainstream, water quality, and the PNPCA for notification and consultation for proposed projects. The PNPCA was adopted in November 2003 and sets out three processes (Notification, Prior Consultation, or Specific Agreement) that member countries must undertake for proposed development projects depending on the type of river (mainstream or tributary), the season (dry or wet), and the scope of water use (inter-basin or intra-basin) (MRC, 2016). Proposals for hydropower dams on the river's tributaries would only require the proposing member country to notify the MRC and member states of the project's details before commencing its development. However, the mainstream dams fall under Prior Consultation, which is a six-month process of formal consultations and technical evaluation, "where notified member states have an opportunity to assess any potential transboundary impact on eco-systems and livelihoods, and to recommend measures to address those issues before water is used" (MRC, 2016: 3).

In accordance with the 1995 Mekong Agreement, Prior Consultation is "neither a right to veto the use nor unilateral right to use water by any riparian without taking into account other riparians' rights" (MRC, 1995: 3). Rather, the role of the MRC is to facilitate a technical evaluation of a proposed project's risks. Prior Consultation is a six-month process involving the submission of impact assessment documents by the proposing member state and an evaluation of the documents by member states. The MRC Secretariat plays a key technical role to "review, analyse and provide technical advice to the MRC JC" and to "provide available technical support for evaluation" (MRC, 2003b: 8). One of the major outputs of Prior Consultation is a technical review of the proposed projects, which sets out to "determine compliance with the MRC procedures on flow regime and water quality, assesses any possible impact on environment and livelihoods, and suggests measures to address those concerns" (MRC, 2016: 4). Here, the MRC's mandate is characterised by technical duties and facilitation carried out by the MRC Secretariat, JC, and NMCs.

Although Prior Consultation is primarily a process of technical evaluation, the public participation component within Prior Consultation is a small but important one. The PNPCA states that one of the responsibilities of the NMCs is to "facilitate any consultations, presentations, evaluation and site visited as requested by the MRC JC for the proposed use" (MRC, 2003b: 7). When Prior Consultation was initiated for the first time for the Xayaburi Dam, the PNPCA Joint Committee Working Group (JCWG) decided at their first meeting on 26 October 2010 to incorporate public participation into the process (MRC, n.d.(e)). This inclusion, rather than being a 'normal and essential' process, appeared to be driven by pressure from internal stakeholders (Vietnamese officials), external stakeholders (civil society), and MRC Secretariat personnel (Whitehead, 2011). Notwithstanding the limited influence of public participation on decision making, its inclusion was nonetheless significant for civil society and local Mekong communities who could directly ask questions about the project where few public platforms for doing so existed,[4] and where project documents were made available on the MRC's website. This is significant as the project information would otherwise be mediated behind the scenes through the MRC's series of PNPCA working groups and has opened up public avenues through which the MRC's limited mandate has been spotlighted. Through this process, external stakeholders ranging from local communities to researchers have also emerged (Geheb et al., 2015).

The contestations arising around the PNPCA reflect the territorial pluralism associated with hydrosocial territories that are differentially defined by multiple stakeholder groups. Although hydropower dams have already been constructed on the river's mainstream in China and on its tributaries in the lower Mekong basin, the construction of hydropower dams on lower Mekong mainstream will significantly reorder the hydrosocial territories of the Mekong by fragmenting, enclosing, and parcelling these transboundary commons (Grundy-Warr, 2017; Miller, 2020). The two existing Lao mainstream dams, the Xayaburi and the Don Sahong, tested the limits of the PNPCA. The Xayaburi Dam is located 150 km downstream from Luang Prabang and has an installed capacity of 1,285 MW (ICEM, 2010). The USD 3.5 billion dam was financed by six Thai banks, developed by Thai construction company CH. Karnchang, and 95 per cent of the electricity generated from the dam is sold to the Thai state utility, the Electricity Generating Authority of Thailand (EPPO, 2016). Being the first mainstream dam to undergo Prior Consultation, and the first dam on the lower Mekong mainstream, the Xayaburi Dam was fiercely contested as it was likely to set a precedent for mainstream dam construction on the lower Mekong mainstream (Osborne, 2013).

The Don Sahong Dam is built across the Hou Sahong channel in southern Laos, one of more than 10 rocky channels through which the Mekong passes before entering Cambodia. The Hou Sahong is the only channel through which fish can

[4] Interview with international NGO representative in Thailand, 13 February 2017.

easily pass through all year round, including the dry season (Baird, 2011). It has an installed capacity of 260 MW and was developed by Malaysian company Mega First Corporation Berhad (ICEM, 2010). The Lao government initially only notified the MRC of the proposed project, given that the dam would not span the whole width of the Mekong mainstream (Boer et al., 2015). However, Laos came under pressure from MRC member states to submit the project under its prior consultation process (Barron, 2013), which Laos reluctantly agreed to do during the MRC Council meeting on 26 June 2014 (Boer et al., 2015). Both the Xayaburi and the Don Sahong are run-of-river dams, which differ from large reservoir dams in that the inflow of water is technically the same as the output flow. This led the Lao government to describe the Xayaburi as a 'transparent' dam (Geheb et al., 2014).

Both the Xayaburi and Don Sahong PNPCAs followed similar trajectories, where the MRC Secretariat's technical review of the key project documents noted that there were significant gaps in the project information submitted, and that the transboundary impacts of the projects were not adequately addressed (MRC Secretariat, 2011, 2015). Strong opposition also arose from civil society, community representatives, scientists, and researchers over the poor quality of the EIAs and the potentially unknown but damaging transboundary impacts on food and livelihood security. However, the Lao government declared the Prior Consultation to be complete after the six-month consultation period, ignoring calls from member countries, particularly downstream states Cambodia and Vietnam, which suggested that it should be extended for further impact assessments and wider consultations to be carried out. With the MRC JC unable to come to a consensus on how to proceed, the final decision was deferred and elevated to the ministerial level of the MRC Council. However, the MRC Council was also unable to reach a decision and Laos proceeded with the ground-breaking and construction of the dams, reflecting the Lao government's determination to proceed with the dams despite the differing views that persisted among MRC member states. Both dams are now operational. These cases revealed weaknesses in the MRC's mandate that allowed Laos to proceed unimpeded, stemming from the ambiguity of the PNPCA's legal status and its guidelines, which were not ratified at the national level and thus were not legally binding as a 'treaty' according to international law (Rieu-Clarke, 2015). The 1995 Mekong Agreement and PNPCA thus fall into the realm of 'soft' rather than 'hard' law (Boer et al., 2015). In contrast to 'hard' law where legal norms "are relatively clear and binding", such as in legislation and treaties, 'soft' law is relatively weak and subject to the discretion of decision makers, although they may still exert some 'compliance pull' (Boer et al., 2015: 45). Through the Xayaburi and Don Sahong PNPCAs, member countries sought to impose and test various interpretations of the procedures within the limits of the MRC's mandate. Rieu-Clarke (2015) identified several areas of ambiguity and contention in his analysis of the

Xayaburi PNPCA, relating to disagreements over what 'timely' notification meant, divergent expectations around availability and relevance of information required to evaluate the proposed project, and the difficulty of adhering to the six-month timeframe. The lack of dispute resolution mechanisms also reflected the lack of authority vested in the MRC Council by member governments, despite the 1995 Mekong Agreement's statement that MRC Council members should be empowered to make decisions on behalf of their governments.

Given that the 1995 Mekong Agreement and the MRC are structured to prioritise the interests and authority of member states, the MRC is "normally considered a governed rather than governing organisation"—its responsibilities are largely implementation as opposed to regulatory (Hirsch, 2011a: 158). While it is acknowledged that the MRC is not: (a) a supranational organisation with regulatory power; (b) an organisation that can make decisions or intervene in its own right; or (c) one that is directly accountable to the broader public, it has been argued that there are still nonetheless grey areas in which the MRC can exert authority (Lee and Scurrah, 2009). These areas relate to the allowable extent of public engagement and accountability, its role as a proactive knowledge-based river basin organisation, its responsiveness to knowledge requests by the public and its capacity to influence development decisions in the Mekong River basin. For example, Suhardiman et al. (2015) have noted that although the MRC operates in a constrained political environment, it had commissioned and used a Strategic Environmental Assessment (SEA) on hydropower development on the Mekong River's mainstream (see ICEM, 2010) as a way of opening up political space and discussion on dams to the wider public, even though the SEA was never officially endorsed by member states.

Criticisms and perceptions of the MRC's ineffectiveness can be said to revolve around the MRC's varying governance mandate to engage in the development of the lower Mekong's water resources on the river's mainstream, tributaries, and lands (Dore and Lazarus, 2009). Due to the ambiguity of the 1995 Mekong Agreement (Boer et al., 2015), the role and mandate of the MRC has become subject to multiple understandings and interpretations by multiple stakeholders (Lee and Scurrah, 2009). This is underlain by a tension between two broad groups of stakeholders. The first group comprises mostly 'external' stakeholders who advocate for what the MRC *should* do, particularly from the perspectives of DPs, civil society, and academics who are concerned with protecting the ecological health of the river basin and livelihoods of rural and vulnerable communities at a regional scale. The second group comprises mostly 'internal' stakeholders, largely MRC staff and member country officials, who are more concerned with what the MRC *can* do in accordance with the 1995 Mekong Agreement. The case of mainstream hydropower development, in particular, has revealed the tensions between these stakeholders. This is representative of the territorial pluralism associated with competing hydrosocial territories put forth by multiple stakeholders.

Through the 1995 Mekong Agreement, the Mekong River Basin as a hydrosocial territory is rendered discursively, spatially, and temporally fragmented. The Agreement reimagines the Mekong River basin as a governable space relating to the creation of powerful discourses that envision the Mekong River primarily as a watercourse rather than a complex socio-ecological system (Bakker, 1999; Sneddon and Fox, 2006). This is reminiscent of James Scott's (1998) concept of state simplification, in the sense that the river has been rendered legible according to state-centric reasoning (Käkönen and Hirsch, 2009). The PNPCA defines the boundaries of this governable space through its technical evaluation of proposed hydropower dams, and, in line with Rose's (1999: 33) conceptualisation, renders visible the entities within them, and assembles information about its included component parts. The limitations of the MRC's mandate enabled member state narratives to take precedence over that of the MRC, underlain by a dichotomy where "[i]n the case of the MRC, such narratives emphasise the environment and sustainable development, in the case of the Lao state, modernity and national development" (Suhardiman and Geheb, 2022: 321).

Creation of a powershed: A fragmented watershed

It is also important to consider the regional dynamics of the power grid whereby Lao hydropower is exported to MRC member states, turning water as a common pool resource into an economic commodity (Tran and Suhardiman, 2020). The PNPCA has reinforced the grounds for a hydrosocial territory increasingly defined by a 'powershed' that is undergirded by a fragmented watershed. Prior Consultation has been carried out in quick succession for four more Lao mainstream dams: the Pak Beng, Pak Lay, Luang Prabang, and Sanakham dams (see Table 3.2). In 2022, it was reported that the Lao government would soon submit the Phou Ngoy dam to the MRC for review. Due to the prior controversies, the MRC Secretariat and the NMC Secretariats have identified the need to manage stakeholder expectations by clarifying the procedures, timeline, and parameters associated with the PNPCA process.[5]

There have also been efforts to address post-PNPCA concerns. After the conclusion of the Pak Beng PNPCA, the MRC JC requested that the MRC Secretariat support the preparation of Joint Action Plans (JAPs), which would "provide mechanisms for ongoing feedback, data exchange, and knowledge sharing between Lao PDR… and the MRC concerning the ongoing design, construction, and operation" of the Pak Beng hydropower project (UNESCAP, 2018: 4). The JAP would be implemented for

[5] Interviews with Thai NMC Secretariat officials in Thailand, 28 February 2017, 27 March 2017; interview with Cambodia NMC Secretariat official, 14 September 2017.

Table 3.2: List of Prior Consultations to date

Period of Prior Consultation	Hydropower Dam Project	Proposing country
Oct 2010 – Apr 2011	Xayaburi Dam	Laos
Jun 2014 – Jan 2015	Don Sahong Dam	Laos
Dec 2016 – Jun 2017	Pak Beng Dam	Laos
Jun 2018 – Apr 2019	Pak Lay Dam	Laos
Oct 2019 – Jun 2020	Luang Prabang Dam	Laos
Jul 2020 – Jan 2022	Sanakham Dam	Laos

all subsequent projects submitted through the PNPCA. The MRC Secretariat noted that the JAP "is not a continuation of Prior Consultation, and it relies on the good faith cooperation in disclosures by the notifying Country" (MRC, 2020a: 2), which potentially raises questions about the extent to which the Lao government will engage meaningfully in the JAP.

The MRC also developed the Joint Environmental Monitoring (JEM) Programme for Mekong Mainstream Hydropower Projects, which was piloted on the Xayaburi and Don Sahong dams in 2020 with support from German Development Cooperation. The JEM Programme would monitor the environmental impacts of these hydropower dams and assess the effectiveness of mitigation measures taken, which would in turn support the JAP (MRC, 2020a). These post-consultation processes reflected what the former MRC Secretariat CEO An Pich Hatda noted as a shift from characterisations of the proposed mainstream dams as "transparent" in the sense that they would not have any impacts on the mainstream, towards an approach premised upon "transparency" in the implementation of the Prior Consultation and post-Prior Consultation processes. This would be "focused on building a better hydropower project rather than a 'transparent' hydropower project", which also "recognized that some transboundary harmful impacts cannot be eliminated" (Hatda, 2020: 2). Since the Pak Beng PNPCA, the MRC also released a 'Council Study' in 2017 which assessed the cumulative impacts of hydropower development in the Mekong River basin, the MRC Hydropower Mitigation Guidelines in 2019, its Sustainable Hydropower Development Strategy in 2021, and the Guidelines for Transboundary Environmental Impact Assessment in the Lower Mekong River Basin in 2023. Many of these also fell under the umbrella of the MRC's Initiative on Sustainable Hydropower which was launched in 2009.

Despite these developments, subsequent iterations of the PNPCA continued to be plagued by criticisms around the poor quality of EIAs, the lack of timely dissemination of project information in local languages, and the highly technical nature of the consultations. The uneven quality of community-level consultations in Thailand, Cambodia, and Vietnam can also be attributed to differing power dynamics inherent in state–society relations, and the resources, networks, and interests of their respective NMC Secretariats (Yong, 2021; Yong and Gillespie, 2022). Among certain segments of community and civil society representatives, a sense of frustration and futility arose as questions of compensation and accountability could not be answered within the technical parameters of the PNPCA. This led to a boycott of the Pak Lay PNPCA by the Save the Mekong Coalition, a transnational activism network. The JAP and JEM are likely to emphasise the MRC Secretariat's technical role in environmental monitoring, and do not appear to provide for public avenues of information dissemination and feedback that Prior Consultation does. Rather, the JAP states that 'external stakeholders' such as DPs, civil society, academics, and other groups will be "kept informed of progress with the implementation of the JAP through the MRC's regular engagements" (MRC, 2019b: 3). The hydrosocial territories of the Mekong have therefore not only been physically fragmented through hydropower development, but also discursively fragmented in terms of whose knowledge counts in managing the transboundary commons.

The hydrosocial territories that come into being through the PNPCA have also legitimised a problematic narrative of 'sustainable' hydropower development, drawing parallels with the Nam Theun 2 dam which has been hailed as a 'model' for sustainable hydropower in Laos (Middleton, 2018). As a result of the PNPCA, the Xayaburi Dam's main consultant, the Finnish engineering company Pöyry, spent about USD 200 million on further fish research and a redesign of the fish passage facilities (Cronin and Weatherby, 2015). This occurred despite warnings from the SEA that fish passes were "not a realistic measure to mitigate the impact of mainstream dams on mainstream fish migrations" (Baran, 2010: 115), reflecting how hydropower development proponents have been 'producing uncertainty' around the risks of such projects (Whitington, 2018). The Xayaburi Dam has been characterised by its proponents as being 'state-of-the-art' (Geheb et al., 2015). However, the true 'sustainability' of these projects is doubtful. The JEM programme's preliminary monitoring showed substantial decreases to sediment load downstream of the dam since 2018, and a 40-60 per cent reduction in biodiversity around the dam (MRC, 2022). Downstream communities have also blamed the Xayaburi Dam's operations for turning the Mekong River an unnatural shade of aquamarine blue in 2019 due to drought and a lack of sediment (Board, 2020). The transformation of the Mekong Region into a powershed through the discourse of sustainable hydropower has thus led to the creation of novel ecologies within the hydrosocial territories of the river basin.

The MRC has attempted to address the limitations and weakness of its mandate through the mechanisms discussed above, but it remains to be seen how much influence the MRC Secretariat's work will have among MRC member states and project developers. For example, the Council Study was not referenced in the EIAs submitted for the Sanakham Dam, which relied on out-of-date information and did not account for changes resulting from the operations of the Xayaburi Dam (MRC, 2021b). Nonetheless, the technical frameworks implemented by the MRC run the risk of rendering the terms of the debate anti-political (Käkönen and Hirsch, 2009) despite the inherent politics of knowledge, power dynamics, and political economy that drives hydropower development in the region (Yong, 2023). The river basin as a hydrosocial territory is discursively fragmented through competing narratives (Geheb et al., 2015) and an ontological politics of water (Middleton, 2022) occurring between a wide range of state and nonstate actors. Today, this hydrosocial territory is also increasingly spatially fragmented through the physical barriers posed by hydropower dams, and temporally fragmented through MRC procedures that did not extend to the construction and operational phases of hydropower dam construction (Yong, 2021).

The relevance of a mandate: China and the Lancang-Mekong River

The territorial pluralism that shapes the Mekong hydrosocial territory is also increasingly influenced by China. The absence of China from the MRC has also always been a key criticism and missing link in the MRC's ability to enact a 'whole-of-basin' approach towards river basin management. China has 'steadfastly declined' to be a full member of the MRC, in part because China perceived the MRC's focus on water resources management to be narrow, and because it did not want to be bound by the predefined rules set out by the 1995 Mekong Agreement, which it had no part in negotiating (Biba, 2018). As the most upstream country, China has constructed 12 hydropower dams on the Lancang since the 1990s. The Xiaowan and Nuozhadu dams are the two largest storage dams that make up about 85 per cent of the Lancang cascade's storage and thus control water availability in Northern Thailand and Laos (Vu et al., 2022). This cascade has been considered as part of a 'domino' effect that has facilitated the Lao government's justification for building its own cascade of dams along the mainstream in the lower Mekong basin (Hirsch, 2011b). In Northern Thailand's Chiang Rai Province, Thai communities living along the Mekong River have complained about unexpected water fluctuations resulting from water releases from Chinese dams having particularly negative impacts on dry-season livelihoods when the unexpected rise in water levels washes away riverbank gardens and boats (Yong, 2020). Prolonged periods of drought in the river basin, for example from 2015 to 2016 and 2019 to 2020 have also highlighted China's potential chokehold over downstream states.

The establishment of the Lancang-Mekong Cooperation (LMC) mechanism in 2016 has raised questions about the continued relevance of the MRC's mandate, particularly in relation to whether the LMC will supplement or replace the MRC in water resources management (Biba, 2018). There are key differences between the two: while the 1995 Mekong Agreement sets out obligations for member states to ensure the reasonable and equitable use of the Mekong River's waters, the LMC is a broader cooperation mechanism comprising economic, political, cultural, and water resources cooperation (Middleton and Allouche, 2016). The Mekong's hydrosocial territory is also home to overlapping regional cooperation mechanisms, of which the MRC is only one of many: the GMS; Mekong-Ganga Cooperation mechanism (India, Myanmar, MRC member states); Ayeyawady-Chao Phraya-Mekong Economic Cooperation (ACMECS) (Myanmar, MRC member states); Mekong-Japan Cooperation; Mekong-Republic of Korea (ROK) Cooperation; and the Lower Mekong Initiative (LMI) spearheaded by the US in 2009. The LMC's establishment represents a changing regional order and was set up for reasons including China's desire to increase its influence over the region on its own terms, to navigate great power dynamics in the region, and to advance major Chinese initiatives such as the Asian Infrastructure Investment Bank and the Belt and Road Initiative (Biba, 2018). Biba (2018: 638) also argues that "China's current LMC strategy appears to essentially replicate the country's former approach" which prioritises the economic benefits to be obtained from and beyond the Mekong River over ecological benefits to the river itself.

Middleton and Devlaeminck (2021: 249) have argued against an oversimplified narrative of China's unilateralism in dam building, pointing out that China has demonstrated "limited acts of reciprocity [with Mekong downstream countries] in navigation, hydropower, flood and drought management, and economic regionalisation". However, these acts of reciprocity are also questioned, particularly with concerns around China's financial institutions' overinvestment in the Lao hydropower sector and allegations of its debt trap diplomacy (Barney and Souksakoun, 2021). Laksmana (2022: 2) notes that the Mekong Region's strategic importance has grown, giving rise to questions around whether the Mekong River would become a 'great power geopolitical playground' and the 'next South China Sea' flashpoint in Southeast Asia, although commentators note that this is unlikely considering the growth of sub-regional cooperation mechanisms (Pich, 2022) and the shared cooperation and benefits between Mekong countries in the energy and water sector (Yong, 2022). Rogers et al. (2023: 3) note that with growing interest in China's 'increased prominence in global development', such as its extra-territorial investments and role in transboundary water governance, there is an "urgent need to interrogate notions of China as a regional hydro-hegemon controlling upstream waters and as a monolithic authoritarian state", an idea which would otherwise create "fixed assumptions rather than an understanding of Asia's dynamic and varied hydropolitics". Rather, Chinese initiatives and infrastructural development are also characterised by fragmentation, decentralised entities and processes, and unintended consequences (Rogers et al., 2023: 3), for example in considering the key role that Yunnan plays as an "energy conduit" between China and Southeast Asia (Ptak and Hommel, 2016).

Biba (2018) acknowledges that the LMC "certainly has the potential to actually implement sustainable water resources management for the Mekong", contingent on the LMC's relationship with the MRC, the extent to which non-state actors will be involved, and the type of hegemon that China wants to be. This has become increasingly pertinent in the realm of data sharing and transparency in relation to China's hydropower dam operations, prolonged drought, and water shortages in the river basin. It has become clearer that Mekong riparian countries will have to recognise how environmental benefits to the Mekong underpin benefit sharing in the economic and political realms, as Biba (2018) has suggested. The Mekong River basin experienced severe droughts and record-low water levels in 2019 and 2020, generating strong concerns across the river basin (Keovilignavong et al., 2023). In April 2020, the US-based consultancy Eyes on Earth released a study funded by the Lower Mekong Initiative (Ketelsen et al., 2020), which concluded that the "severe lack of water in the Lower Mekong during the wet season of 2019 is largely influenced by the restriction of water flowing from the Upper Mekong during that time", attributable to China's dam operations (Basist and Williams, 2020: 18). These results generated significant regional debate and media attention and were used by US government representatives to criticise China's actions, which in turn elicited reactions from China's diplomats (Middleton et al., 2021).

The study's claims and methodology were also scrutinised by the MRC (2020b), academics (Kallio and Fallon, 2020), and the Australia-Mekong Partnership for Environmental Resources and Energy Systems (AMPERES), which also questioned the ways in which the study's conclusions were used for geopolitical purposes (Ketelsen et al., 2020). This has led to suggestions that the drought was 'hydropoliticised' (Middleton et al., 2021) and that science was being politicised around the Eyes on Earth study (Grünwald et al., 2021). Underlying these contested claims was a continued lack of data on China's dam operations. Notwithstanding the controversies associated with the Eyes on Earth study, the US-based think tank Stimson Center's Mekong Dam Monitor—which is run in partnership with Eyes on Earth—has done what the MRC perhaps cannot unilaterally do due to the political sensitivities around data sharing, such as providing near-real-time tracking of dam operations, and issuing early warning alerts relating to China's upstream dam releases (Eyler et al., 2023).

China has shared wet season (June–October) water level and rainfall data from two hydrological stations since 2002, but it was only in October 2020 that China signed an agreement with the MRC to provide year-round hydrological data, which then-MRC Secretariat CEO An Pich Hatda described as a "landmark in the history of China-MRC cooperation" (MRC, 2020c). Since then, a joint MRC-LMC technical study has been carried out, which acknowledged that the "[n]ear real-time sharing of storage levels and hydropower operations will also be critical to support operational models and adaptive management of the basin" and recommended enhanced data and information sharing, along with the coordinated management of water resources (LMWRCC-MRC, 2023). These recommendations were endorsed by senior officials from the six Mekong countries in September 2023, which the MRC stated would "pave the way for better information sharing about reservoir operations on the Mekong's water flow, and water release as well as practical solutions to mitigate and adapt to impacts" (MRC, 2023).

The LMC reflects China's attempt to exert influence on the reshaping of the Mekong River as a hydrosocial territory, by replicating its former approach of emphasising economic and political benefits to be gained from and beyond the Mekong River. However, the events of the 2019 and 2020 drought highlight how drastic changes to the biophysical conditions of the river can also play a driving role in reshaping this hydrosocial territory in material, discursive, and geopolitical terms. While the MRC's mandate has been limited because of China's absence from the MRC, it may have found renewed relevance in its mandate around data sharing and transparency, albeit under circumstances largely beyond its control and realm of responsibility where external pressures generated by the debate around the Eyes on Earth study appear to have played a role in 'incentivising' China to engage in such activities.

There has been an increase in cooperative activities between the MRC and China through the LMC, particularly in the areas of hydrological data sharing and study activities, but it remains to be seen if information sharing around reservoir operations across the Mekong River Basin will come to fruition. This is in part because this will be dependent on coordination between the hydrosocial networks of the river basin, comprising a range of heterogeneous actors, including hydropower dam operators. For example, Ptak and Hommel's (2016) study points out there is a need to pay closer attention to the coordination that takes place between China's local, provincial, and central governments, power entities such as hydropower operators and grid operators, and China's ongoing negotiation of relationships with entities such as the GMS and downstream Mekong countries.

Conclusion

This chapter has sought to understand how the hydrosocial territories of the Mekong River Basin have been shaped through three phases which shed light on different forms of 'regionness' and the ways in which the MRC's mandate has been situated in them. The first phase traced the MRC's history and demonstrated how the lower Mekong River basin came to be defined in relation to the upper Mekong

River basin through the formation of the MC, IMC, and MRC. Significant geopolitical and geoeconomic changes across the river basin, and contested negotiations, led to a weakened MRC mandate as expressed in the 1995 Mekong Agreement, revealing how sovereign and national interests could take precedence over any notion of a 'region' that emphasised the ecological health and integrity of the Mekong River basin as a watershed.

The second phase identified in this paper considered the MRC's limited mandate as a key component of a critical institutionalism approach to understanding the make-up of the Mekong's hydrosocial territories. Here, a focus on the PNPCA served as a gateway to understand how the Mekong's watershed is spatially, temporally, and discursively fragmented, giving way to the creation of a 'powershed'. The ways in which mainstream hydropower development have been contested by a wide range of stakeholders highlights the territorial pluralism and ontological politics that characterise hydrosocial territories. While the PNPCA initially gave space for the voices of external stakeholders to emerge, it has become increasingly clear that the MRC's technical and limited mandate has facilitated a narrative of 'sustainable' hydropower to emerge as a part of the river basin's hydrosocial territories.

Finally, a consideration of the emergence of the LMC amidst increasing geostrategic competition and a politicisation of science around the 2019-2020 drought highlights how China and the US have tried to assert themselves in the region, contributing to a changing regional order. The MRC has once again been thrust into the spotlight given the contestations around data sharing and transparency, which may lead to new ways of understanding and shaping the hydrosocial territories of the Mekong River basin. However, it remains to be seen if it will be able to proactively assert its role in river basin planning, or if the MRC will continue to be subject to reacting to external regional forces beyond its control.

References

The Don Sahong Dam: potential impacts on regional fish migrations, livelihoods, and human health. *Critical Asian Studies* 43(2): 211–235, https://doi.org/10.1080/14672715.2011.570567.

Bakker, K. 1999. The politics of hydropower: developing the Mekong. *Political Geography* 18(2): 209–232, https://doi.org/10.1016/S0962-6298(98)00085-7.

Baran, E. 2010. Mekong fisheries and mainstream dams. Fisheries sections of the Strategic Environmental Assessment of hydropower on the Mekong mainstream prepared for the Mekong River Commission, http://www.researchgate.net/profile/Eric_Baran/publication/259529258_Fisheries_sections_in_the_Strategic_Environmental_Assessment_of_Mekong_mainstream_dams/links/00b4952c6424a465fe000000.pdf (accessed 2 May 2022).

Barney, K. and Souksakoun, K. 2021. Credit crunch: Chinese infrastructure lending and Lao sovereign debt. *Asia & the Pacific Policy Studies* 8(1): 94–113, https://doi.org/10.1002/app5.318.

Barron, L. 2013. Nations unite against dam. *Phnom Penh Post*. 2 December 2013, https://www.phnompenhpost.com/national/nations-unite-against-dam (accessed 1 January 2019).

Basist, A. and Williams, C. 2020. Monitoring the quantity of water flowing through the upper Mekong Basin under natural (unimpeded) conditions. Bangkok: Sustainable Infrastructure Partnership, https://558353b6-da87-4596-a181-b1f20782dd18.filesusr.com/ugd/bae95b_0e0f87104dc8482b99ec91601d853122.pdf?index=true (accessed 17 November 2022).

Biba, S. 2018. China's 'old' and 'new' Mekong River politics: the Lancang-Mekong Cooperation from a comparative benefit-sharing perspective. *Water International* 43(5): 622–641, https://doi.org/10.1080/02508060.2018.1474610.

Billo, E. and Mountz, A. 2016. For institutional ethnography: geographical approaches to institutions and the everyday. *Progress in Human Geography* 40(2): 199–220, https://doi.org/10.1177/0309132515572269.

Board, J. 2020. 'The colour is blue': strange changes to Mekong River as hydropower dams and climate change make their mark. *Channel News Asia*. 12 January 2020, https://www.channelnewsasia.com/asia/mekong-river-colour-xayaburi-dam-laos-hydroelectric-power-1338616 (accessed 8 March 2024).

Boelens, R., Hoogesteger, J., Swyngedouw, E., Vos, J. and Wester, P. 2016. Hydrosocial territories: a political ecology perspective. *Water International* 41(1): 1–14, https://doi.org/10.1080/02508060.2016.1134898.

Boer, B., Hirsch, P., Johns, F., Saul, B. and Scurrah, N. 2015. *The Mekong: a socio-legal approach to river basin development*. Abingdon: Routledge.

Browder, G. 2000. An analysis of the negotiations for the 1995 Mekong Agreement. *International Negotiation* 5(2): 237–261, https://doi.org/10.1163/15718060020848758.

Browder, G. and Ortolano, L. 2000. The evolution of an international water resources management regime in the Mekong River Basin. *Natural Resources Journal* 40(3): 499–532, https://digitalrepository.unm.edu/nrj/vol40/iss3/2.

Cleaver, F. and de Koning, J. 2015. Furthering critical institutionalism. *International Journal of the Commons* 9(1): 1–18, https://thecommonsjournal.org/articles/10.18352/ijc.605.

Cronin, R. and Weatherby, C. 2015. Letters from the Mekong: time for a new narrative on Mekong hydropower. Washington D.C.: The Henry L. Stimson Center, https://www.stimson.org/sites/default/files/file-attachments/Letters_from_the_Mekong_Oct_2015.pdf (accessed 6 May 2024).

Dore, J. and Lazarus, K. 2009. De-marginalizing the Mekong River Commission. In Molle, F., Foran, T. and Kakonen, M. (eds.), *Contested waterscapes in the Mekong Region: Hydropower, Livelihoods and Governance*, 358–381. London: Earthscan.

Dore, J. and Lebel, L. 2010. Deliberation and scale in Mekong Region water governance. *Environmental Management* 46(1): 60–80, https://doi.org/10.1007/s00267-010-9527-x.

EPPO (Energy Policy and Planning Office). 2016. Xayaburi Hydroelectric Power Project. Bangkok: EPPO, Ministry of Energy, https://www.eppo.go.th/index.php/en/energy-information-services/xayaburi-hydroelectric-power-project (accessed 6 February 2024).

Eyler, B., Basist, A., Kwan, R., Weatherby, C. and Williams, C. 2023. Mekong dam monitor annual report: 2022–2023. Washington D.C., USA: The Stimson Center, https://www.stimson.org/wp-content/uploads/2023/12/Mekong-Dam-Monitor-Report_Web.pdf (accessed 26 March 2024).

Flaminio, S., Rouillé-Kielo, G. and Le Visage, S. 2022. Waterscapes and hydrosocial territories: thinking space in political ecologies of water. *Progress in Environmental Geography* 1(1–4): 33–57, https://doi.org/10.1177/27539687221106796.

Geheb, K., West, N. and Matthews, N. 2015. The invisible dam: hydropower and its narration in the Lao People's Democratic Republic. In Matthews, N. and Geheb, K. (eds.), *Hydropower development in the Mekong Region: political, socio-economic and environmental perspectives*, 101–126. London: Earthscan.

Gerlak, A.K. and Haefner, A. 2017. Riparianization of the Mekong River Commission. *Water International* 42(7): 893–902, https://doi.org/10.1080/02508060.2017.1376267.

Glassman, J. 2010. *Bounding the Mekong: The Asian Development Bank, China and Thailand*. Honolulu (HI): University of Hawaii Press.

Grundy-Warr, C. 2017. B/ordering nature and biophysical geopolitics: a response to Hirsch. *Political Geography* 58: 131–135, https://doi.org/10.1016/j.polgeo.2016.10.006.

Grünwald, R., Feng, Y. and Wang, W. 2021. Politicization of science in the Lancang–Mekong Basin: the Eyes on Earth Study. *International Journal of Water Resources Development* 39(2): 184–210, https://doi.org/10.1080/07900627.2021.1990025.

Hatda, A.P. 2020. From transparent to transparency: the evolution of the Mekong River Commission's prior consultation process. *Asia Pacific Bulletin* No. 527, 8 October 2020. Honolulu: East-West Center, https://www.eastwestcenter.org/publications/transparent-transparency-the-evolution-the-mekong-river-commissions-prior-consultation (accessed 12 February 2024).

Hirsch, P. 2011a. IWRM as a participatory governance framework for the Mekong River Basin? In Öjendal, J., Hansson, S. and Hellberg, S. (eds.), *Politics and development in a transboundary watershed*, 155–170. Dordrecht: Springer Netherlands, https://doi.org/10.1007/978-94-007-0476-3_8.

Hirsch, P. 2011b. China and the cascading geopolitics of lower Mekong dams. *The Asia-Pacific Journal* 9(20): 1–4, https://apjjf.org/2011/9/20/philip-hirsch/3529/article.

Hirsch, P. 2016. The shifting regional geopolitics of Mekong dams. *Political Geography* 51: 63–74, https://doi.org/10.1016/j.polgeo.2015.12.004.

ICEM (International Centre for Environmental Management). 2010. Strategic environmental assessment on hydropower on the Mekong Mainstream. Hanoi: ICEM, http://www.icem.com.au/documents/envassessment/mrc_sea_hp/SEA_Final_Report_Oct_2010.pdf (accessed 10 September 2015).

Jacobs, J.W. 1995. Mekong Committee history and lessons for river basin development. *The Geographical Journal* 161(2): 135–148, https://doi.org/10.2307/3059971.

Käkönen, M. and Hirsch, P. 2009. The anti-politics of Mekong knowledge production. In Molle, F., Foran, T. and Kakonen, M. (eds.), *Contested waterscapes in the Mekong Region: hydropower, livelihoods and governance*, 333–355, London: Earthscan.

Kallio, M. and Fallon, A. 2020. Are China's dams on the Mekong causing downstream drought? The importance of scientific debate. CRITICAL NATURE Blog, 28 April 2020. Bangkok, Center for Social Development Studies, Chulalongkorn University, https://www.csds-chula.org/publications/2020/4/28/critical-nature-are-chinas-dams-on-the-mekong-causing-downstream-drought-the-importance-of-scientific-debate (accessed 18 March 2024).

Keovilignavong, O., Nguyen, T.H. and Hirsch, P. 2023. Reviewing the causes of Mekong drought before and during 2019–20. *International Journal of Water Resources Development* 39(1): 155–175, https://doi.org/10.1080/07900627.2021.1967112.

Ketelsen, T., Sawdon, J. and Räsänen, T. 2020. Monitoring the quality of water flowing through the upper Mekong Basin under natural (unimpeded conditions): Rapid Review, https://www.amperes.com.au/library/2020/4/amperes-commentary-eyes-on-earth (accessed 13 March 2024).

Kittikhoun, A. 2023. The 2nd State of the Mekong Address, https://www.youtube.com/watch?v=LuyIK-70e1A (accessed 24 June 2026).

Kossov, I. and Lay, S. 2016. Donors slash funding for MRC. *Phnom Penh Post* 14 January 2016, https://www.phnompenhpost.com/national/donors-slash-funding-mrc (accessed 19 June 2018).

Kuus, M. 2017. Critical Geopolitics. In *Oxford Research Encyclopedia of International Studies*, https://doi.org/10.1093/acrefore/9780190846626.013.137.

Laksmana, E.A. 2022. Will the Mekong River be the next South China Sea? *Counterpoint Southeast Asia* (3): 1-3, https://lkyspp.nus.edu.sg/docs/default-source/cag/csa-3---1.pdf?sfvrsn=97b9300a_2 (accessed 7 July 2024).

Lee, G. and Scurrah, N. 2009. Power and responsibility: the Mekong River Commission and lower Mekong mainstream dams. Sydney: Australian Mekong Resource Centre, University of Sydney and Oxfam Australia, http://sydney.edu.au/mekong/documents/power_and_responsibility_fullreport_2009.pdf (accessed 26 October 2015).

LMWRCC-MRC (Lancang-Mekong Water Resources Cooperation Center and Mekong River Commission). 2023. Technical Report – Phase 1 of the Joint study on the Changing Patterns of Hydrological Conditions of the Lancang-Mekong River Basin and Adaptation Strategies. Beijing and Vientiane: LMC Water Center and MRC Secretariat, https://doi.org/10.52107/mrc.bdozhw.

Middleton, C. 2016. Water, rivers and dams. In Hirsch, P. (ed.), *Routledge Handbook of the Environment in Southeast Asia*, 204–223. London: Taylor & Francis Group.

Middleton, C. 2018. Branding dams: Nam Theun 2 and its role in producing the discourse of "sustainable hydropower." In Shoemaker, B. and Robichaud, W. (eds.), *Dead in the water: global lessons from the World Bank's model hydropower project in Laos*, 271–292. Madison (WI): University of Wisconsin Press.

Middleton, C. 2022. The political ecology of large hydropower dams in the Mekong basin: a comprehensive review. *Water Alternatives* 15(2): 251–289, https://www.water-alternatives.org/index.php/alldoc/articles/vol15/v15issue2/668-a15-2-10/file (accessed 22 January 2023).

Middleton, C. and Allouche, J. 2016. Watershed or powershed? Critical hydropolitics, China and the 'Lancang-Mekong Cooperation Framework.' *The International Spectator* 51(3): 100–117, https://doi.org/10.1080/03932729.2016.1209385.

Middleton, C. and Devlaeminck, D.J. 2021. Reciprocity in practice: the hydropolitics of equitable and reasonable utilization in the Lancang-Mekong basin. *International Environmental Agreements: Politics, Law and Economics* 21(2): 235–253, https://doi.org/10.1007/s10784-020-09511-6.

Middleton, C., Garcia, J. and Foran, T. 2009. Old and new hydropower players in the Mekong Region. In Molle, F., Foran, T. and Käkönen, M. (eds.), *Contested waterscapes in the Mekong Region: hydropower, livelihoods and governance*, 23–54. London: Earthscan.

Middleton, C., Widyasari, A., Manorom, K., Devlaeminck, D.J. and Intralawan, A. 2021. Strengthening Water Diplomacy Through Water Data Sharing and Inclusive Evidence-Based Transboundary Governance. Bangkok and Phnom Penh: Center for Social Development Studies, Faculty of Political Science, Chulalongkorn University, and Cambodia Development Research Institute, https://www.csds-chula.org/s/CSDS-and-CDRI-2021-Water-Diplomacy-Research-Report.pdf (accessed 18 March 2024).

Miller, M.A. 2020. B/ordering the environmental commons. *Progress in Human Geography* 44(3): 473–491, https://doi.org/10.1177/0309132519837814.

Mills-Novoa, M., Boelens, R., Hoogesteger, J. and Vos, J. 2020. Governmentalities, hydrosocial territories & recognition politics: The making of objects and subjects for climate change adaptation in Ecuador. *Geoforum* 115: 90–101, https://doi.org/10.1016/j.geoforum.2020.06.024.

Mirumachi, N. (2011) Domestic water policy implications on international transboundary water development: A case study of Thailand. In Öjendal, J., Hansson, S. and Hellberg, S. (eds.), *Politics and development in a transboundary watershed*, 83–100. Dordrecht: Springer Netherlands.

MRC (Mekong River Commission). 1995. Agreement on the cooperation for the sustainable development of the Mekong River Basin, http://www.mrcmekong.org/assets/Publications/policies/agreement-Apr95.pdf (accessed 10 July 2017).

MRC (Mekong River Commission). 2003a. State of the basin report 2003. Vientiane: MRC, http://www.mekonginfo.org/assets/midocs/0001489-inland-waters-state-of-the-basin-report-2003.pdf (accessed 23 May 2024).

MRC (Mekong River Commission). 2003b. Procedures for notification, prior consultation and agreement, https://tile.loc.gov/storage-services/service/gdc/gdcovop/2017357526/2017357526.pdf (accessed 12 May 2024).

MRC (Mekong River Commission). 2016. Procedural rules for Mekong water diplomacy: Procedures for Notification, Prior Consultation and Agreement (PNPCA), http://www.mrcmekong.org/assets/Publications/PNPCA-brochure-11th-design-final.pdf (accessed 2 May 2024).

MRC (Mekong River Commission). 2019a. State of the basin report 2018. Vientiane: MRC, https://www.mrcmekong.org/assets/Publications/SOBR-v8_Final-for-web.pdf (accessed 24 May 2024).

MRC (Mekong River Commission). 2019b. Joint action plan for the implementation of the statement on the prior consultation process for the Pak Beng Hydropower Project. Vientiane: MRC, https://doi.org/10.52107/mrc.ajg6bo (accessed 14 February 2024).

MRC (Mekong River Commission). 2020a. Joint action plan for the implementation of the statement on the prior consultation process of the Luang Prabang Hydropower Project. Vientiane: MRC, https://www.mrcmekong.org/assets/Publications/Joint-Action-Plan-on-Luang-Prabang-project.pdf (accessed 14 February 2024).

MRC (Mekong River Commission). 2020b. Understanding the Mekong River's hydrological conditions: a brief commentary note on the "Monitoring the quality of water flowing through the upper Mekong basin under natural (unimpeded) conditions" study by Alan Basist and Claude Williams (2020). Vientiane: MRC Secretariat, https://www.mrcmekong.org/wp-content/uploads/2024/08/Understanding-Mekong-River-hydrological-conditions_2020.pdf (accessed 18 March 2024).

MRC (Mekong River Commission). 2020c. China to provide the Mekong River Commission with year-round water data, https://www.mrcmekong.org/news-and-events/news/china-to-provide-the-mekong-river-commission-with-year-round-water-data/ (accessed 18 March 2024).

MRC (Mekong River Commission). 2021a. The integrated water resources management-based basin development strategy for the Mekong River Basin 2021-2030 & MRC Strategic Plan 2021-2025. Vientiane: MRC Secretariat, https://www.mrcmekong.org/assets/Publications/BDS-2021-2030-and-MRC-SP-2021-2025.pdf (accessed 25 May 2024).

MRC (Mekong River Commission). 2021b. Prior consultation process for the Sanakham Hydropower Project: summary of the final draft of the technical review report. Vientiane: Mekong River Commission, https://www.mrcmekong.org/assets/Consultations/Sanakham/Technical-Review-Report-Final-Draft/Summary-of-Final-Draft-of-the-Technical-Review-Report-EN.pdf (accessed 8 March 2024).

MRC (Mekong River Commission). 2022. Joint environmental monitoring programme at two Mekong mainstream dams: the Don Sahong and Xayaburi hydropower projects. Vientiane: MRC, https://doi.org/10.52107/mrc.aqrs70 (accessed 22 February 2024).

MRC (Mekong River Commission). 2023. Six Mekong-Lancang countries reached pact on more dam operations data sharing. *Mekong River Commission*, https://www.mrcmekong.org/news-and-events/news/pr-11092023/ (accessed 19 March 2024).

MRC (Mekong River Commission). n.d.(a) Fast facts on the Mekong River Commission, https://www.mrcsummit.org/assets/documents/Press-Docs/f700ef0c10/Fast-facts-about-the-MRC_EN.PDF (accessed 2 May 2024).

MRC (Mekong River Commission). n.d.(b) MRC functions, https://www.mrcmekong.org/our-work/functions/ (accessed 30 January 2024).

MRC (Mekong River Commission). n.d.(c) Governance and organisational structure, https://www.mrcmekong.org/about/mrc/governance-and-organisational-structure/ (accessed 30 January 2024).

MRC (Mekong River Commission). n.d.(e) Xayaburi Hydropower Project, http://www.mrcmekong.org/topics/pnpca-prior-consultation/xayaburi-hydropower-project-prior-consultation-process/ (accessed 6 October 2018).

MRC Secretariat. 2011. Procedures for notification, prior consultation and agreement (PNPCA) proposed Xayaburi Dam Project – Mekong River: prior consultation project review report, https://archive.internationalrivers.org/sites/default/files/attached-files/mrc_technical_review_of_xayaburi_march_2011.pdf (accessed 8 May 2018).

MRC Secretariat. 2015. Technical review report on prior consultation for the proposed Don Sahong Hydropower Project, http://www.mrcmekong.org/assets/Publications/Consultations/Don-Sahong/Technical-Review-Report-DSHPP-040315.pdf (accessed 1 January 2019).

Osborne, M.E. 2013. Xayaburi Dam's domino effect? *The Interpreter* 23 January 2013, https://www.lowyinstitute.org/the-interpreter/xayaburi-dam-s-domino-effect (accessed 11 June 2024).

Philo, C. and Parr, H. 2000. Institutional geographies: introductory remarks. *Geoforum* 31(4): 513–521, https://doi.org/10.1016/S0016-7185(00)00018-X.

Pich, C. 2022. Proliferation of cooperative mechanisms hinder conflict. *Counterpoint Southeast Asia* (3): 7–9, https://lkyspp.nus.edu.sg/docs/default-source/cag/csa-3---3.pdf?sfvrsn=1db9300a_0 (accessed 20 November 2023).

Ptak, T. and Hommel, D. 2016. The trans-political nature of southwest China's energy conduit, Yunnan Province. *Geopolitics* 21(3): 556–578, https://doi.org/10.1080/14650045.2015.1114471.

Rieu-Clarke, A. 2015. Notification and consultation procedures under the Mekong Agreement: Insights from the Xayaburi controversy. *Asian Journal of International Law; Singapore* 5(1): 143–175, https://doi.org/10.1017/S2044251314000022.

Rogers, S., Fung, Z., Lamb, V., Gamble, R., Wilmsen, B., Wu, F. and Han, X. 2023. Beyond state politics in Asia's transboundary rivers: Revisiting two decades of critical hydropolitics. *Geography Compass* 17(4): e12685, https://doi.org/10.1111/gec3.12685.

Rose, N. 1999. *Powers of freedom: reframing political thought*. Cambridge: Cambridge University Press.

Scott, J. 1998. *Seeing like a state: how certain schemes to improve the human condition have failed*. New Haven (CT): Yale University Press.

Sneddon, C. and Fox, C. 2006. Rethinking transboundary waters: a critical hydropolitics of the Mekong basin. *Political Geography* 25(2): 181–202, https://doi.org/10.1016/j.polgeo.2005.11.002.

Stimson Center. 2020. Mekong infrastructure tracker dashboard. Washington D.C.: Stimson Center, https://www.stimson.org/2020/mekong-infrastructure-tracker-tool/ (accessed 25 March 2024).

Stimson Center. 2024. Mekong mainstream dams. Washington D.C.: Stimson Center, https://www.stimson.org/2020/mekong-mainstream-dams/ (accessed 26 March 2024).

Suhardiman, D. and Geheb, K. 2022. Participation and politics in transboundary hydropower development: the case of the Pak Beng dam in Laos. *Environmental Policy and Governance* 32(4): 320–330, https://doi.org/10.1002/eet.1974.

Suhardiman, D., Giordano, M. and Molle, F. 2015. Between interests and worldviews: the narrow path of the Mekong River Commission. *Environment and Planning C: Government and Policy* 33(1): 199–217, https://doi.org/10.1068/c11191.

Swyngedouw, E. 1999. Modernity and hybridity: nature, *regeneracionismo*, and the production of the Spanish waterscape, 1890–1930. *Annals of the Association of American Geographers* 89(3): 443–465, https://doi.org/10.1111/0004-5608.00157.

Tran, T.A. and Suhardiman, D. 2020. Laos' hydropower development and cross-border power trade in the Lower Mekong Basin: A discourse analysis. *Asia Pacific Viewpoint* 61(2): 219–235, https://doi.org/10.1111/apv.12269.

UNESCAP (United Nations Economic and Social Commission for Asia and the Pacific). 2018. Seventy-fourth Session, Bangkok 11-16 May, 2018. Item 3(d) of the provisional agenda: review of issues pertinent to the subsidiary structure of the Commission, including the work of the regional institutions: environment and development – Report of the Mekong River Commission, https://www.unescap.org/sites/default/d8files/event-documents/E74_INF2E.pdf (accessed 12 January 2024).

Vu, D.T., Dang, T.D., Galelli, S. and Hossain, F. 2022. Satellite observations reveal 13 years of reservoir filling strategies, operating rules, and hydrological alterations in the Upper Mekong River basin. *Hydrology and Earth System Sciences* 26(9): 2345–2364, https://doi.org/10.5194/hess-26-2345-2022.

Whitehead, I. 2011. Closed channels or open waters? Mekong dams, transboundary law and the Xayaburi prior consultation process. Master's Thesis. Sydney: The University of Sydney.

Whitington, J. 2018. *Anthropogenic rivers: the production of uncertainty in Lao Hydropower*. Ithica: Cornell University Press.

Wright, S. 2016. Mekong effort fails after years of lavish foreign funding. *AP News*. 18 October 2016, https://apnews.com/international-news-general-news-31978ed8726449dca8ba47c628161 37a (accessed 30 January 2024).

Yong, M.L. 2020. Reclaiming community spaces in the Mekong River transboundary commons: shifting territorialities in Chiang Khong, Thailand. *Asia Pacific Viewpoint* 61(2): 203–218, https://doi.org/10.1111/apv.12257.

Yong, M.L. 2021. Transboundary environmental publics and hydropower governance in the Mekong River Basin: a contested politics of place, scale and temporality. *Environmental Policy and Governance*, https://doi.org/10.1002/eet.1973.

Yong, M.L. 2022. Different geopolitical and material realities. *Counterpoint Southeast Asia* #3: 4–6, https://lkyspp.nus.edu.sg/docs/default-source/cag/csa-3---2.pdf?sfvrsn=b4b9300a_0 (accessed 11 June 2024).

Yong, M.L. 2023. Hybrid governance, environmental justice, and hydropower development in the Mekong transboundary commons. *WIREs Water* 10(5): e1665, https://doi.org/10.1002/wat2.1665.

Yong, M.L. and Gillespie, J. 2022. Towards relational geometries of public participation and hydropower governance in the Lower Mekong River Basin. *Political Geography* 99: 102773, https://doi.org/10.1016/j.polgeo.2022.102773.

CHAPTER 4

Revisiting the interplay between water management and adaptation in the Vietnamese Mekong Delta: A political ecology perspective

Thong Anh Tran[1,2] and Le Anh Tuan[3]

Abstract

The Vietnamese Mekong Delta has undergone dramatic environmental change driven by climate change, upstream hydropower development, and extensive state investment in water infrastructure. Drawing on political ecology, desk reviews, and key informant interviews, this chapter examines the political dynamics at the interface of water management and adaptation. We argue that state-led infrastructure development has compelled 'forced adaptation': farmers must respond to environmental change driven not by climate alone but by water infrastructure operations. This reveals the power asymmetries between state-led water management and farmer-led adaptation, demonstrating how the state constructs modernity and legitimacy through infrastructure mandates. We conclude that addressing these contestations requires a political understanding of climate-development contexts and state–society relations.

Keywords: adaptation, climate change, hydropower development, political ecology, Vietnamese Mekong Delta, water management

Introduction

The Vietnamese Mekong Delta (VMD) lies at the mouth of the Mekong River. The 'hydropower boom' in the upper stretches of the river has resulted in critical transboundary issues in downstream countries, e.g., recurring depletion to water flows, sediment, fisheries, biodiversity, and other aquatic ecosystems. In addition to these drivers, the accelerating effects of climate change and associated

[1] Fenner School of Environment and Society, College of Systems and Society, Australian National University, Canberra, Australia.

[2] Mekong Institute, Can Tho University, Can Tho City, Vietnam.

[3] College of Environment and Natural Resources, Can Tho University, Can Tho City, Vietnam.

natural phenomena, such as El Niño, working in tandem with the legacy of state-led 'modernistic' water engineering development across the VMD have dramatically transformed its agro-ecological landscape (Tran and Tortajada, 2022).

The historical development processes of the VMD manifest the interwoven yet contested nature of water management and adaptation that takes place thereafter. Essentially, water management has been understood as a technocratic, state-led approach of building and expanding water infrastructural systems for agricultural production (Biggs et al., 2009), demonstrated by the widespread construction of dykes and canals to control floods in the floodplains (Tong, 2015) and to implement 'freshening coastal zones' policies to retain freshwater and prevent saline intrusion into the coastal delta (Tran et al., 2022). While these state-led efforts have yielded a certain level of success in terms of expanding the VMD's cultivated area, realities suggest that farmers contest such policies by engaging in various forms of 'everyday politics' (Gorman, 2023; Tran and Pittock, 2024). While such evidence is salient, however, there is a lack of a political ecological understanding of the *in-situ* politics operating behind the relational dynamics and decision-making processes of water management and adaptation in the broader climate and development context of the lower Mekong Basin.

Existing theories of political ecology foreground queries around capitalist accumulation processes, with a particular focus on the human domination of the environment (Zinzani, 2023). Political ecology seeks to understand the complex relations between nature and society through the analysis of forms of access and control over resources and their implications for sustainable livelihoods (Watts, 2000: 257). It critically analyses dominant accounts of environmental change, while exploring alternatives, and adaptation in the face of mismanagement and exploitation (Tran, 2020). This chapter uses political ecology to unravel the relatively overlooked politics that emerge from the interplay of water management and adaptation in the VMD (Table 4.1), illuminating the contested state–society relations operating in the authoritarian Vietnamese regime of environmental governance (Ortmann, 2017).

Water management tends to become more politicised in times of environmental crisis (Popartan and Ungureanu, 2022). Given the climate-development complexities of the Mekong Region, water management and adaptation are high on the Vietnamese government's political agenda in seeking a climate-resilient and sustainably developed delta as directed by Resolution 120 (GoV, 2017).[4] At the local

[4] Resolution 120 is a Vietnamese government policy intended to guide climate-resilient and sustainable development trajectories in the VMD. The document includes an overarching development agenda in relation to the government's planning strategies, with long-term visions, objectives, overall solutions, and concrete tasks to improve economic development while prioritising adaptation to climate change (GoV, 2017).

Table 4.1: Political relations between water management and adaptation in the VMD.

Categories	Water management	Adaptation
Aims	Flood control, water retention, prevention of saline intrusion.	Rural and agricultural modernisation. Diversifying farm-based livelihoods for enhanced household income. Accommodating environmental change.
Approach	Hard measures (technical approaches), such as infrastructure systems.	Soft approaches (non-technical approaches), such as changes to farming systems.
Descriptions	Water management schemes (dykes, sluices) that impose a physical geographical boundary in the floodplains or coastal environments.	Livelihood activities exercised within the geographical boundary determined by water infrastructure systems.
Key actors	Government agencies	Farmers
Power relations	Led by the state which holds predominant power in decision making (e.g., investment in water engineering schemes in water control).	Led by farming communities who are often marginalised and under-represented in decision-making processes, particularly in water scheme development.
Impacts	Leads to ecological change inside/outside water infrastructure systems.	Affected by ecological change due to water control systems.

scale, water management has close political relations to adaptation. This chapter employs a qualitative case study approach guided by Baxter and Jack (2008) to investigate how these political dynamics occurred in the context of hydraulic development processes in the delta, taking An Giang (in the floodplains) and Ben Tre (in the coastal zones) as case studies (see Fig. 4.1). It argues that the state's investment in water infrastructure for water expulsion in the floodplains (via the construction of lateral drainage canals to the West Sea (Gulf of Thailand)) and freshening the salinity zones in the coastal plains (through the building of the Ba Lai sluice scheme) has triggered 'forced' adaptation. This evolving predicament forces farmers to deal with undesirable water consequences (e.g., water scarcity) under the immediate impacts of the infrastructure operations.

The chapter contributes to advancing a nuanced understanding of the socio-political context along the confluence of these two responses to climate change and the sustainability crisis in the VMD. It portrays a relatively under-explored dimension of power asymmetries between state and farming communities. The chapter emphasises that while state exercises its dominant power and political mandates over water and/through infrastructure systems, farming communities are placed in

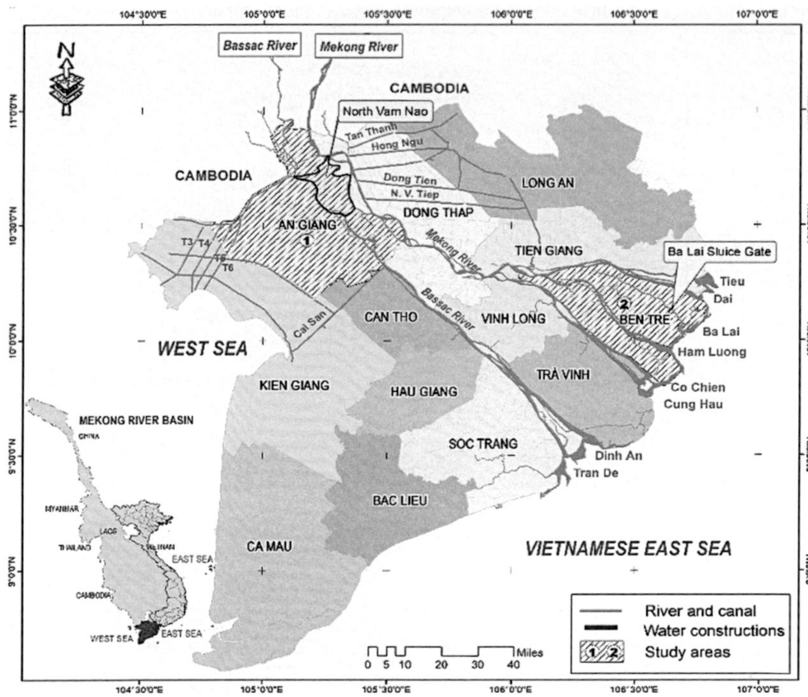

Figure 4.1. The VMD and the case study areas (shaded): (1) An Giang Province in the floodplains and (2) Ben Tre Province in the coastal zone (map credit: Dung Duc Tran).

a subordinate position. Even so, farmers were restrained from accessing resources or other sources of capital to enable their adaptation. They also lack the power to exercise their rights and interests in the pursuit of profitable livelihoods (e.g., shrimp farming) that fit the local ecosystem conditions and are forcibly bound to adopt new farming practices, i.e., freshwater-based farming systems such as the fruit, rice, or sugarcane demanded by the local government.

Case study contexts

Two provinces situated in the two distinct ecological zones of the VMD were selected for the study: An Giang in the floodplains and Ben Tre in the coastal zones (Fig. 4.1). These case study areas are representative of the key water management policies implemented in the delta, and how they shape the adaptation practices of farming communities. For example, in the floodplains of the Long Xuyen Quadrangle (including An Giang Province), a series of lateral canals, including T4, T5, and T6, played a pivotal role in draining off excess floodwaters to the West Sea, while assisting in reclaiming acid sulphate soils for rice production (Lich, 2022; van Aalst et al., 2023). In Ben Tre Province, the Ba Lai scheme, which turned the Ba Lai River into a reservoir, enables the year-round retention of freshwater (Tessier and Anh, 2022), allowing for diversified agricultural systems, minimising water scarcity, and preventing saline intrusion into freshwater zones during the dry season.

An Giang is annually exposed to high flooding (Tuan et al., 2007). Agricultural and aquacultural production constitute the economic backbone for this province, where rice is the main staple crop. The density of dyke systems (low and high dykes) allows local inhabitants to cultivate multiple rice crops per year. The seasonally flooded environments also enable rural farmers, especially the poor, to make a living based on the collection of wild capture fisheries and other aquatic resources (Tran and James, 2017) (Table 4.2). Here, annual floods occur from July to December (Tuan et al., 2007). The province has, however, experienced recurring waterflow disruptions, which come as a coupled consequence of climate change and upstream hydropower operations (Tran and Tortajada, 2022), posing significant challenges to floodplain communities and their water-based livelihoods.

Ben Tre is a coastal province located in the eastern part of the VMD, bordering the East Sea (South China Sea) to the southeast. With the coast's length at about 65 km, the province is a favourably endowed environment, allowing for the growth of brackish and saline aquaculture and marine fisheries. In recent years, tidal effects combined with other climatic processes have triggered saline intrusion, which is perceived to be increasingly severe (Apel et al., 2020). The Ba Lai sluice, built on the Ba Lai River, attempts to arrest this phenomenon and retain freshwater for agricultural production and household consumption (Hoang et al., 2009). The Ba Lai sluice divides the two sides of the river (Binh Dai and Ba Tri districts) into freshwater zones (upper reaches) and estuary zones (lower reaches), each with distinct agro-ecological systems. Freshwater agro-commodities are mainly cultivated in the freshwater zones, including bananas, coconuts, and rice, whereas brackish and saltwater aquaculture dominate the estuary zones, including integrated farming systems (shrimp-rice), intensive/extensive shrimp farming, blood cockles, as well as salt production.

State-led infrastructure-based water management

In the VMD, the state has prioritised infrastructure-based water management in both the floodplains and the coastal zones. The floodplains, which were previously exposed to high floods, have been transformed into 'dykescapes' to meet national food security policies (Biggs et al., 2009). Efforts to control floods enabled the rapid development of water infrastructure systems. A wide range of low and high dykes

Table 4.2: Description of the case study areas with details collected before the national government's reorganisation of administrative units. Sources: GSO (2020); Tran and Cook (2024); Tessier and Anh (2022).

Categories	Case study areas	
	An Giang Province	Ben Tre Province
Geographical locations	In the floodplains (directly connected to the Mekong River).	In the coastal zones (adjacent to the East Sea (South China Sea)).
Total land area (km^2)	3,537	2,395
Population (people)	2,164,200	1,268,200
Population density (peoples/km^2)	612	530
Water infrastructure development in the study areas	Lateral canals (built in the late 1990s) connected to the West Sea to allow the drainage of floodwater in the flood season.	The Ba Lai sluice scheme (built in 2000 and operated in 2002) devoted to the year-round retention of freshwater (up to 90 million m^3) and prevention of saline intrusion. Forming a socially and environmentally demarcated boundary between freshwater zones (upstream of the Ba Lai River) and estuary zones (downstream).
Main livelihoods	Agricultural and aquacultural commodities (rice, upland crops, freshwater fish, giant freshwater prawn (*Macrobrachium rosenbergii*)), seasonal resource collection (wild capture fisheries).	Rice, fruits, cash crops, and brackish and saline polycultures (integrated farming systems (shrimp-rice)), intensive/extensive shrimp, blood cockles, salt).
Major water-related risks	Disruptions of floodwater regimes, water scarcity.	Saline intrusion, water scarcity.

and canals were built, which aimed to prevent the entry of floods into rice fields as well as drain off excess floodwater to the West Sea. Local adaptation narratives in An Giang indicated that farmers were collectively involved in building low dykes (also known as 'August dykes') to prevent early flooding into their rice fields, enabling them to harvest a second rice crop (the Summer-Autumn crop) (Tong, 2015). In the late 1990s, lateral canals that connected the floodplains of the Long Xuyen

Quadrangle to the West Sea were built, confirming the state's emphasis on technical and engineered solutions to the perceived challenges of flooding (Lich, 2022).

The investment in water infrastructure systems in the coastal zones began from the launch of the 'rice everywhere' national campaign, of which the 'freshening coastal zones' campaign was a part. These campaigns started in the Ca Mau Peninsula in the late 1990s (Cosslett and Cosslett, 2014; Tran et al., 2022) and aimed to convert the areas of permanent salinity into freshwater zones, making them available for rice production (Benedikter, 2014). The soils in these new freshwater environments remained, however, prone to high acid sulphate concentrations, leading to low cultivated rice productivity (Gowing et al., 2006). The construction of the Ba Lai sluice system in Ben Tre Province, which was seen as the second phase of the 'freshening coastal zones' campaign, aimed to retain freshwater and safeguard the province from the accelerating rate of saline intrusion (Tran et al., 2022). Realities suggested that these water infrastructure systems pose critical constraints to *in-situ* water environments, making it difficult for farming communities to adapt. The following section will elaborate on how these domains are entangled in contested relations.

Political relations between water management and adaptation

There exist unbalanced political dynamics between water management and adaptation as evidenced in the contested relations between state and farming communities across governance scales. At the national scale, water management conventionally sets priority on flood control and the provision of freshwater for agricultural development (Waibel et al., 2012). At the local scale, this is demonstrated by the large-scale state investment in water infrastructure systems, such as dykes and canals in An Giang to control floods as well as sluices and reservoirs in Ben Tre to retain freshwater and prevent saline intrusion.

Government rhetoric highlights the detrimental impacts of climate change and upstream hydropower development (including the Central Highlands of Vietnam), and how such infrastructural investments safeguard local livelihoods. This engineering, however, has constrained farmers' abilities to adapt their livelihoods to environmental change. These constraints include farmers' struggles in shifting crop systems and gaining access to resources, such as capital, technology, and knowledge to meet their needs.

Water infrastructure is an 'iconic' feature in the state's intention to develop. In other words, it puts on display the ways in which the state seeks dominant power and legitimacy to translate its ideologies into actionable policies/mandates to meet its needs (e.g., exploiting the VMD via environmental control). This is, however, grounded in the state's simplistic understandings of the environment as implied in Scott's (1998) conceptualisation of simplification and legibility, in which knowledge is overwhelmingly technocratic and dominates decision-making processes. It also demonstrates an effort by the government in homogenising and simplifying pre-existing diverse social and natural environments (Harris, 2017). This, as a result, contributes to environmental transformation. For example, the water expulsion policy implemented in the Long Xuyen Quadrangle in the late 1990s resulted in a suite of lateral canals delivering excess floodwaters to the West Sea (Lich, 2022). These efforts have, however, led to mixed results. While the construction of the canals contributed to converting acid sulphate soils into cultivable land areas and provided an unprecedented opportunity to produce multiple rice crops in the high flood zones of the VMD, they also caused water scarcity and saline intrusion during the dry season. In the Ben Tre case study, the local government's construction of the Ba Lai scheme was also grounded in a similar ideology that converted the river into a reservoir to hold freshwater for agricultural development (Tran et al., 2022). This contributed to transforming the coastal landscapes that depend on freshwater from the reservoir for livelihood development.

Power asymmetries between state-led water management and farmer-led adaptation were evident in the study areas (Table 4.3). Government agencies, especially the provincial ones, dominated water management decision making (Tran, 2020). Given the advantage of their power, government agencies are in a better position to mobilise resources and capacity for water infrastructure investments. The building of lateral canals in the Long Xuyen Quadrangle was, for example, the collective effort of national and provincial government agencies (Lich, 2022). In the domain of adaptation, farmers hold a subordinate position, in which they lack the political power and show their passive submission to state water management policies. This skewed power relation constitutes an underlying socio-political boundary in relation to government water management that marginalises farmers, especially the poor, from participating in rural development policies (Tran and Rodella, 2019).

The development of water infrastructure has had detrimental effects on local environments, triggering state–society contestations. Besides the combined effects of climate change and upstream hydropower development, emerging water challenges in the VMD are also ascribed to the side-effects of water infrastructure systems (Tran and Tortajada, 2022). The floodwater expulsion policy in the Long Xuyen Quadrangle was initially justified as making space for rice production (Benedikter, 2014), but ultimately ended up driving water scarcity (Tran and Cook, 2023). Disrupted waterflows have become increasingly frequent (Le, 2020), leading to large-scale water shortages. The weakening of the Mekong flood pulse has implications for coastal zones, which are affected by saline intrusion in the dry season (Binh et al., 2020). The construction of the Ba Lai sluice system enabled local inhabitants to cultivate multiple crops and curbed saline intrusion. The local government, however, also restricted access to groundwater (brackish water) for shrimp farming in the freshwater zones. Shrimp farmers complained that the scheme had dispossessed them of their means of livelihood, arguing that brackish/saline resources could also bring economic returns as high as those from freshwater (Tran et al., 2022). Through the lens of water justice (Sultana, 2018), these comparative outcomes beg the question whether the technical water management approach assumes the ecological responsibility that needs to be fairly distributed to those who are not in agreement with the state-led water development.

As indicated, the political relations between water management and adaptation represent competing values and interests between state and farming communities. The state has maintained its dominant power by manipulating water to fulfil its political mandate through water infrastructural development. At the same time, however, farming communities are forced to adapt their livelihoods to environmental change—in part, created by these infrastructural interventions.

Shouldering the impacts: How farmers enact 'forced' adaptation

In addition to the adverse effects of climate change, the state's infrastructural interventions have posed numerous challenges to the aquatic ecosystems on which farming communities depend for their livelihoods. The transformation of these ecosystems alters the nature of adaptation, forcing farmers to adjust their livelihoods. Trapped in this forced adaptation context, farmers are subject to complying with the 'new rules of the game', which at times they are unlikely to adhere to or negotiate with. These 'new rules of the games' are concerned with social and environmental complexities arising from the new human–environment relationships in the wake of the infrastructural interventions.

Rural communities in the Mekong floodplains have experienced critical constraints in dealing with dwindling floodwater regimes. While smallholder farmers were able to shift their farming systems to minimise these impacts, poor farmers were faced with losing seasonal employment opportunities (e.g., wild fish capture). Others have formed rural 'communities of practice', which serve as an informal interaction space for participants who shared similar cultural and social practices and concerns, and who seek to learn and generate knowledge (Wenger, 2000). In this space, farmers learned with one another (mostly from within their bonding network systems including relatives, friends, and neighbours) and self-organised other collective activities to reduce environmental impacts (Tran et al., 2018). In tandem with other *in-situ* development impacts (e.g., land use change driven by rural agricultural mechanisation), attempts to secure livelihoods prompted poor farmers to migrate to

Table 4.3: Political relations between water management and adaptation in the case study areas.

Case study areas	An Giang		Ben Tre	
Categories	Water management	Adaptation	Water management	Adaptation
Structural systems in focus	Lateral canals connecting floodplains (Long Xuyen Quadrangle) to the West Sea.	Farmers can cultivate multiple (2-3) rice crops per year.	The Ba Lai sluice system built on the Ba Lai River (freshwater reservoir).	Farmers in the freshwater zones diversify farming systems.
Approaches	Flood control; draining off the excess floodwater to the West Sea; reclaiming large areas of acid sulphate soils for rice production.	Flooding communities are forced to adapt to changes in flood regimes and the presence of water infrastructure systems (low/high dykes and canals).	Year-round supply of freshwater for agricultural production; prevention of saline intrusion; piped water supply to households.	Farmers in the freshwater zones learn to live with changes from the building of the Ba Lai sluice (from brackish/saline to freshwater environments).
Descriptions	Spans the Long Xuyen Quadrangle, including An Giang Province.	Flood-based livelihoods change into water infrastructure-based livelihood practices.	Ben Tre Province bordering the East Sea.	Farmers' livelihoods in both freshwater and estuary zones are shaped by the closing and opening of the Ba Lai scheme sluice gates.
Key actors	National and local government agencies.	Farmers self-organise livelihoods and adaptation to change (rice cultivation).	National and local government agencies.	Farmers in the upper zones cultivate freshwater crops (citrus fruits, rice, bananas, coconuts); farmers in the lower zones are affected by freshwater drainage.

Case study areas	An Giang		Ben Tre	
Categories	Water management	Adaptation	Water management	Adaptation
Power relations	National and local governments dominate decision making.	Farmers were marginalised from the planning, management, and operation of canal systems.	National and local governments dominate decision making.	Farmers are not involved in the planning, management, and operation of the Ba Lai sluice system.
Impacts	Contributing to water scarcity under new water conditions (driven by climate change and upstream hydropower development).	Living with 'too much or too little' water. Local inhabitants have experienced disruptions in wild capture fisheries (change in seasonal floods) and declining job opportunities (agricultural mechanisation).	Divided ecological zones between fresh and brackish/saline water zones.	Farmers in the freshwater zones are encouraged to cultivate freshwater crops, while shrimp farming was prohibited due to the likelihood of propagating saline water throughout the freshwater zones.

urban areas in search of employment (Tran, 2019). Migration, in this sense, implied farmers' failure to respond to challenges in their places of origin (Bayrak et al., 2023).

In the context of the Ba Lai sluice, farming communities were also faced with multiple challenges. Those residents in the freshwater zones were encouraged to cultivate freshwater crops, such as citrus fruit, bananas, and rice, which, however, did not bring about desirable economic returns. The recurring market volatility of such agro-commodities has led local farmers to shift to shrimp farming, which was previously practised by the majority of coastal inhabitants before the construction of the Ba Lai scheme. Realities suggested that shrimp farmers engaged in everyday adaptation politics to venture into this livelihood option (Tran and Pittock, 2024), as the pursuit of this farming practice violates water use rules in the freshwater zones (Tran et al., 2022). For instance, shrimp farmers engaged in acts of evasion, dissembling, and deception (Gorman, 2023). Specifically, they built fake wells in an open space to attract government attention while building real ones in the hidden one (Tran and Pittock, 2024). In the downstream (estuary) zones of the Ba Lai scheme, local farmers were exposed to the bi-monthly drainage of (polluted) freshwater from the freshwater zones, threatening the cultivation of blood cockles across the alluvial mudflats of the estuary areas (Tran et al., 2022). As noted by Tessier and Anh (2022), the confluence of freshwater drainage from the upstream zones, combined with the increased level of salt concentration, makes aquacultural production in the downstream zones more challenging.

Overall, the constraints imposed by state-led water management have dramatically eroded the agency among local farmers who were already perceived as marginalised actors by such water infrastructure interventions. Here, adaptation emerges from an increasingly constrained operating space, which is then 'forced', given the lack of alternative livelihood options. The perceived challenges of water scarcity in the floodplains as well as saline intrusion in the coastal zones have placed a disproportionate burden on farmers' shoulders, aggravating their livelihood conditions. In the broader context, this raises a critical question of how the contested relations between water management and adaptation can be negotiated to bring the desired livelihood outcomes to rural farmers.

Working with and for farmers: Ways to navigate contested relations

Addressing contested relations between water management and adaptation calls for policy considerations across governance scales to realise technical and operational flaws as well as institutional arrangements in place within local water infrastructural systems, and to leverage societal capacity in responding to environmental change (Tran and Rodela, 2019). This provides a window of opportunity for relevant stakeholders, particularly local governments and farmers, to engage more closely in deliberative dialogues and shared learning, thereby seeking realistic solutions to address the problems (Smajgl et al., 2015). It is therefore essential to look into the mixed perceptions of decision makers between central and provincial levels and among stakeholders who play a role in water management in the VMD and nationally. At these crossroads, this paper suggests a critical investigation into how water management and adaptation can reinforce each other to enhance farmers' livelihoods in the face of environmental change in the VMD.

Addressing these contested relations suggests the need to centre farmers in water management, where their voices can be articulated and recognised. The successful engagement of local farmers in water management decision making will be translated into productive adaptation, where farmers' livelihoods can be better secured. Engaging local farmers in the planning and development of water infrastructure systems is essential; equal considerations must, however, be given to potential adverse environmental impacts of water management schemes and their effects on the livelihoods of farming communities.

Lessons from the North Vam Nao flood control scheme in An Giang Province—where local communities were actively involved in the planning, construction, and operation of the scheme—are of much relevance here as they can inform pathways for rectifying water management arrangements in the VMD. The establishment of compartment boards provided a space for local communities to meaningfully engage in local decision-making processes related to the scheme's management and operation, enabling more effective responses to various flood conditions (Tran et al., 2020). Its institutional arrangements enabled farmers to confidently communicate with the heads of compartment boards and local government officials in ways that allowed them to articulate their demands and concerns when needed. This has enabled farmers to proactively anticipate the flood impacts and organise their livelihood activities accordingly.

Conclusions

The chapter presents the state's sustained efforts in water infrastructure development, which serves as a backbone for agriculture-based economic growth as well as the response to climate change in the VMD. These have pushed the state to expand water infrastructure investments across the delta over the past few decades. The construction of the lateral canals to the West Sea in the Long Xuyen Quadrangle and the Ba Lai sluice in Ben Tre Province are the products of this political mandate. This type of development intervention has, however, created challenges for farming communities in the case study areas when responding to environmental change.

The power asymmetries in state–society relations are clearly demonstrated through the contestations between state-led water management and farmer-led adaptation. While the state actively exercises its power through political mandates and control over water and infrastructure systems, farming communities are disadvantageously relegated to a subordinate position. Farmers lack the power to exercise their rights and interests in pursuit of preferential livelihoods. Instead, they are forced to confront the environmental consequences of the water infrastructure construction and operations.

Failures to address water management and adaptation equitably will hinder visions towards achieving the resilient and sustainable development sought by Resolution 120 in the delta. Addressing these contested relations therefore needs to place farmers at the centre of water management, ensuring that their voices can be recognised and integrated into decision-making processes, particularly at the local level. This chapter advocates for a holistic approach to collaborative water management that should be built upon a grounded understanding of regional and local climate-development complexities, as well as the aspirations of farming communities directly affected by such policy decisions. More importantly, prioritising water management options, particularly those focused on technical solutions, should not come at the expense of adaptation. Failing to balance these approaches will further risk marginalising farming communities whose livelihoods are already threatened by water challenges in the delta.

References

Apel, H., Khiem, M., Nguyen, H.Q. and To, Q.T. 2020. Brief communication: seasonal prediction of salinity intrusion in the Mekong Delta. *Natural Hazards and Earth System Sciences* 20(6): 1609–1616, https://doi.org/10.5194/nhess-20-1609-2020.

Bayrak, M.M., Tran, H.V., Tran, T.A., Hsu, Y.-Y., Nien, T. and Quynh, D.T.T. 2023. Climate change adaptation responses and human mobility in the Mekong Delta: local perspectives from rural households in An Giang Province, Vietnam. *Humanities and Social Sciences Communications* 10(2023): 344, https://doi.org/10.1057/s41599-023-01817-5.

Baxter, P. and Jack, S. 2008. Qualitative case study methodology: study design and implementation for novice researchers. *The Qualitative Report* 13(4): 544–559, https://doi.org/10.46743/2160-3715/2008.1573.

Benedikter, S. 2014. *The Vietnamese hydrocracy and the Mekong Delta: water resources development from state socialism to bureaucratic capitalism.* Wien: LIT Verlag.

Biggs, D., Miller, F., Chu, T.H. and Molle, F. 2009. The Delta machine: water management in the Vietnamese Mekong Delta in historical and contemporary perspectives. In Molle, F., Foran, T. and Käkönen, M. (eds.), *Contested waterscapes in the Mekong Region: hydropower, livelihoods and governance*, 203–225. London: Earthscan.

Binh, D.V., Kantoush, S.A., Saber, M., Mai, N.P., Maskey, S., Phong, D.T. and Sumi, T. 2020. Long-term alterations of flow regimes of the Mekong River and adaptation strategies for the Vietnamese

Mekong Delta. *Journal of Hydrology: Regional Studies* 32(2020): 100742, https://doi.org/10.1016/j.ejrh.2020.100742.

Cosslett, T.L. and Cosslett, P.D. 2014. *Water Resources and Food Security in the Vietnam Mekong Delta*. New York (NY): Springer.

Gorman, T. 2023. The art of not being freshened: the everyday politics of infrastructure in the Mekong Delta. *Sustainability* 15(2023): 5494, https://doi.org/10.3390/su15065494.

GoV (Government of Vietnam). 2017. *Resolution No. 120/NQ-CP on Sustainable and Climate-Resilient Development of the Vietnamese Mekong Delta*, dated 17 November 2017. Hanoi: Government of Vietnam.

Gowing, J.W., Tuong, T.P., Hoanh, C.T. and Khiem, N.T. 2006. Social and environmental impact of rapid change in the coastal zone of Vietnam: an assessment of sustainability issues. In Hoanh, C.T., Tuong T.P., Gowing, J.W. and Hardy, B. (eds.), *Environment and Livelihoods in Tropical Coastal Zones – Managing Agriculture-Fishery-Aquaculture Conflicts*, 48–60. Cambridge: CABI International.

GSO (General Statistics Office). 2020. *Socio-economic statistical data of 63 provinces and cities*. Hanoi: Statistical Publishing House.

Harris, L.M. 2017. Political ecologies of the state: recent interventions and questions going forward. *Political Geography* 58(2017): 90–92, https://doi.org/10.1016/j.polgeo.2017.03.006.

Hoang, Q.H., Kubo, N., Hoang, N.G. and Tanji, H. 2009. Operation of the Ba Lai irrigation system in the Mekong Delta, Vietnam. *Paddy Water Environment* 7(2009): 123–133, https://doi.org/10.1007/s10333-009-0155-0.

Le, A.T. 2020. Analysis of flood and drought variability in the Mekong River Delta in the last 20 years. *Journal of Science and Technology* 62(11): 22–27, https://vjol.info.vn/index.php/most/article/view/54662/45272.

Lich, D.M. 2022. *Thoát Lũ Ra Biển Tây* [*Drainage of Floodwater out to the West Sea*]. Hanoi: Hong Duc Publishing House.

Ortmann, S. 2017. *Environmental governance in Vietnam: institutional reforms and failures*. Cham: Springer International Publishing.

Popartan, L.A. and Ungureanu, C. 2022. The political ecology of water memory: contending narratives of past hydraulic infrastructures in Barcelona (2015–2021). *Political Geography* 96(2022): 102596, https://doi.org/10.1016/j.polgeo.2022.102596.

Scott, J.C. 1998. *Seeing like a state: how certain schemes to improve the human condition have failed*. New Haven (CT): Yale University Press.

Smajgl, A., Toan, T.Q., Nhan, D.K., Ward, J., Trung, N.H., Tri, L.Q., Tri, V.P.D. and Vu, P.T. 2015. Responding to rising sea levels in the Mekong Delta. *Nature Climate Change* 5(2): 167–174, https://doi.org/10.1038/nclimate2469.

Sultana, F. 2018. Water justice: why it matters and how to achieve it. *Water International* 43(4): 483–493, https://doi.org/10.1080/02508060.2018.1458272.

Tessier, O. and Anh, L.H.H. 2022. Impacts of the Ba Lai dam on production system and ecosystem – Binh Dai District, Ben Tre Province. In: Woillez, M.-N., Espagne, E. (eds.), *The Mekong Delta Emergency: Climate and Environmental Adaptation Strategies to 2050*, 157–179. Final Report – GEMMES Viet Nam Project. Paris: Agence Française de Développement, Paris, https://www.afd.fr/en/ressources/mekong-delta-emergency-climate-and-environmental-adaptation-strategies-2050 (accessed 7 January 2026).

Tong, Y.D. 2015. *A cost-benefit analysis of dyke heightening in the Mekong Delta*. EEPSEA Research Report RR11. Laguna: WorldFish and EEPSEA, https://cdn.cocodoc.com/cocodoc-form-pdf/pdf/344713331--A-COSTBENEFIT-ANALYSIS-OF-DIKE-HEIGHTENING-IN-THE-MEKONG-DELTA-eepsea-.pdf?download=1 (accessed 6 January 2026).

Tran, T.A. 2019. Land use change driven out-migration: evidence from three flood-prone communities in the Vietnamese Mekong Delta. *Land Use Policy* 88(2019): 104157, https://doi.org/10.1016/j.landusepol.2019.104157.

Tran, T.A. 2020. From free to forced adaptation: a political ecology of the 'state-society-flood' nexus in the Vietnamese Mekong Delta. *Asia Pacific Viewpoint* 61(1), 162–182, https://doi.org/10.1111/apv.12241.

Tran, T.A. and Cook, B.R. 2024. Water retention for agricultural resilience in the Vietnamese Mekong Delta: towards integrated 'grey-green' solutions. *International Journal of Water Resources Development* 40(3): 532–553, https://doi.org/10.1080/07900627.2023.2207674.

Tran, T.A. and James, H. 2017. Transformation of household livelihoods in adapting to the impacts of flood control schemes in the Vietnamese Mekong Delta. Water Resources and Rural Development 9(2017): 67–80, https://doi.org/10.1016/j.wrr.2017.04.002.

Tran, T.A., James, H. and Pittock, J. 2018. Social learning through rural communities of practice: empirical evidence from farming households in the Vietnamese Mekong Delta. *Learning, Culture and Social Interaction* 16(2018): 31–44, https://doi.org/10.1016/j.lcsi.2017.11.002.

Tran, T.A. and Pittock, J. 2024. When water policies derail livelihood aspirations: farmers' agency in everyday politics in the Vietnamese Mekong Delta. *Environmental Sociology*, 10(3): 267–278, https://doi.org/10.1080/23251042.2024.2323601.

Tran, T.A. and Tortajada, C. 2022. Responding to transboundary water challenges in the Vietnamese Mekong Delta: in search of institutional fit. *Environmental Policy and Governance* 32(4): 331–347, https://doi.org/10.1002/eet.1980.

Tran, T.A. and Rodela, R. 2019. Integrating farmers' adaptive knowledge into flood management and adaptation policies in the Vietnamese Mekong Delta: a social learning perspective. Global Environmental Change 55(2019): 84–96, https://doi.org/10.1016/j.gloenvcha.2019.02.004.

Tran, T.A., Tran, H.V., Pittock, J. and Cook, B. 2022. Political ecology of freshening the Mekong's coastal delta: narratives of place-based land-use dynamics. *Journal of Land Use Science* 17(1), 471–486, https://doi.org/10.1080/1747423X.2022.2126907.

Tuan, L.A., Hoanh, C.T., Miller, F. and Sinh, B.T. 2007. Flood and salinity management in the Mekong Delta, Vietnam. In Be, T.T., Sinh, B.T. and Miller, F. (eds.), *Challenges to sustainable development in the Mekong Delta: regional and national policy issues and research needs*, 18–68. Bangkok: The Sustainable Mekong Research Network, https://www.researchgate.net/publication/305377337_Flood_and_salinity_management_in_the_Mekong_Delta_Vietnam (accessed 7 January 2026).

van Aalst, M.A., Koomen, E., Tran, D.D., Hoang, H.M., Nguyen, H.Q. and de Groot, H.L.F. 2023. The economic sustainability of rice farming and its influence on farmer decision-making in the upper Mekong delta, Vietnam. *Agricultural Water Management* 276(2023): 108018, https://doi.org/10.1016/j.agwat.2022.108018.

Waibel, G., Benedikter, S., Reis, N., Genschick, S., Loan, N.T.P., Huu, P. C. and Be, T.T. 2012. Water governance under renovation? Concepts and practices of IWRM in the Mekong Delta, Vietnam. In Renaud, F.G. and Kuenzer, C. (eds.), *The Mekong Delta system: interdisciplinary analyses of a river delta*, 167–198. Dordrecht: Springer Environmental Science and Engineering.

Watts, M., 2000. Political ecology. In Sheppard, E., Barnes, T.J. (eds.), *A Companion to economic geography*, 257–274. Malden (MA): Blackwell Publishing Ltd.

Wenger, E. 2000. Communities of practice and social learning systems. *Organization* 7(2): 225–246, https://doi.org/10.1177/135050840072002.

Zinzani, A. 2023. The contested environmental futures of the Dolomites: a political ecology of mountains. *Geographica Helvetica* 78(2): 295–307, https://doi.org/10.5194/gh-78-295-2023.

CHAPTER 5

Hydrocorruption: Water, politics, hydropower, and the state in Laos

Danzig Sopera[1]

Abstract

In this chapter, I explore the role of corruption in Laos's massive hydropower buildout. I argue that corruption is a key—if not dominant—trait in the behaviour of the Politburo of the Lao People's Revolutionary Party (LPRP), affecting its decision making and drawing focus towards the corruption opportunities afforded by hydropower construction. While the Politburo offers rhetoric promoting hydropower and its benefits, I argue that these have not materialised—therefore, the motivations behind the Lao hydropower expansion need to be found elsewhere, in hydrocorruption. I analyse the ways in which hydrocorruption has evolved in Laos, providing examples of both domestic and transnational hydrocorruption praxis.

Keywords: Laos, hydropower development, corruption, political economy

> "[W]e have the potential for hydropower ... so we have to do the dams. That is our choice"
> —Chanthanet Boualapha, Chair of the Lao National Mekong Committee, 2020. (Johnson and Wongcha-Um, 2020)

> "Corruption is a serious bug that eats into a nation, and it will finally topple the nation"
> —Asang Laoly, Deputy Prime Minister of Laos (2005) (High, 2012: 190).

> "... corruption will not go away in Thailand. It is in the system"
> —Thaksin Shinawatra (quoted in Mutebi, 2008).

Introduction

On 31 March 1998, the Theun-Hinboun Hydropower Plant (HPP) was commissioned in the Lao province of Bolikhamxai. With an installed capacity of 200 megawatts (MW), the Theun-Hinboun was the first hydropower plant in Laos to be built under

[1] *Nom de plume.*

a Build-Own-Transfer (BOT) formula, in which its parent company, the Theun-Hinboun Power Company (THPC), was awarded a 25-year concession to build and operate the dam, before transferring it to the Lao state. Coordinated by the Asian Development Bank (ADB), THPC was established through a shareholder agreement between the Government of Laos (GOL) (as the majority shareholder), a Thai firm, and a Norwegian firm. Funding for the dam (about USD 240 million) was raised through loans from the ADB, the Nordic Development Fund, Thai commercial banks, and export credit agencies. Once commissioned, the dam began exports of approximately 1,400 gigawatt hours (GWh) a year to Thailand.

At the time of its commissioning, the Lao economy was in difficulty. With glasnost in full swing, the Soviet aid that Laos had previously relied on was tapering off. In 1998, GDP growth in Laos was 3.97 per cent; the next year, however, it was up at 7.31 per cent—of which THPC was responsible for 7 per cent, a GDP presence that it would maintain for six years (ADB, 1997). For several years, it was also the primary source of foreign exchange for the country. To the GOL, THPC must have seemed like a veritable cash cow. As THPC reminded assembled guests at its 25th Anniversary Celebration in 2023, the firm had contributed more than USD 760 million to state coffers over its quarter century of operations (Vientiane Times, 2023). The Theun-Hinboun demonstrated that the BOT approach provided sufficient guarantees to investors, but it would also reveal the capital intensity of hydro, and the complexity that hydropower construction entailed, as important opportunities for a rent-seeking agenda.

Globally, the water sector is highly vulnerable to corruption, predominantly due to the opportunities presented by its development (Stålgren, 2006). Hydropower development is capital intensive, and the number of contractual, financial, material, and institutional exchanges that must occur to achieve it is spectacular (Transparency International, 2008) (Fig. 5.1). Hydropower's socio-economic footprint is large, and budgets must also accommodate the costs of resettlement and compensation of project-affected people (WCD, 2000). Additionally, hydropower development is susceptible to cost and time overruns (Ansar et al., 2014). When it comes to large hydropower, the decision to proceed is politically highly risky. Once taken, the option to withdraw is limited by the threat of losing face. This characteristic also incentivises corrupt behaviour.

Paul Robbins (2000: 424) defines corruption as "an institutionalised system of nature/society interaction forged from state authority and moulded around local social power through systems of social capital formation". In corrupt countries, it is a constituent part (if not the primary motivation) of how natural resources are governed (Williams and Le Billon, 2017). Yet corruption "is difficult to pin down precisely because it is, by definition, opaque. The complexities of locating corruption are doubled by the potency of the claim as a political weapon" (Elinoff, 2017: 588). In

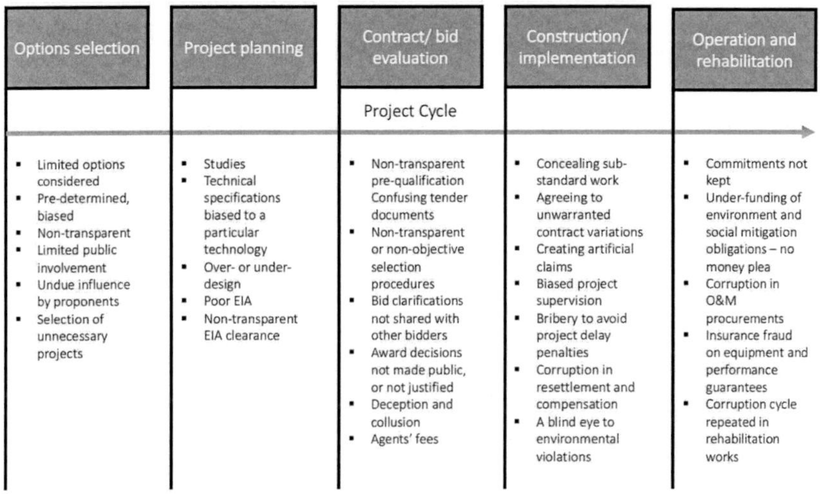

Figure 5.1: Scope and enabling conditions for corruption in various stages of a hydropower project cycle (derived from Haas, 2008).

Laos, "[a]ccusations of corruption have ... become something of a political weapon, and at the central level, such accusations are scandals that can bring projects into disrepute or ostracise personalities" (High, 2012: 104).

In this chapter, I introduce 'hydrocorruption' as an analytical concept for examining how corruption operates systematically within the hydropower sector. The term denotes the entanglement of corruption with hydropower development—not as an aberration but as a structuring logic. Hydrocorruption encompasses the rent-seeking opportunities created by capital-intensive dam construction, the networks of patronage linking political elites to construction firms and financiers, and the institutional arrangements that enable and reproduce corrupt practices. The concept draws attention to how the particular characteristics of hydropower—its scale, complexity, opacity, and path dependencies—make it especially amenable to capture by corrupt interests.

Corruption around infrastructural development in the Mekong Region's river basins is common. There is, of course, little *proof* that it occurs—and this chapter does not set out to prove that any of the public servants named accepted inducements, or that they formulated policy and developed networks to maximise rent-seeking ambitions. More often than not, corruption and the names of those alleged to be corrupt are whispered with information shared on condition of anonymity, and sentences delivered with the preface, 'don't quote me on this, but...'.

This chapter draws on a collage of information sources. Wherever possible, I prioritise the formal research literature, and failing that, the press; several sources were interviewed for a study that I carried out in 2021 (Supera, 2022), and then, there are those 'sources' who declined to be interviewed or anonymously quoted but nevertheless shared small pieces of information fitting into a larger picture. In these cases, reliability was assessed based on the informant's role. Information from those working within relevant ministries or hydropower companies was weighted more heavily than opinion-based information from outsiders.

I focus on Laos, a country where hydropolitics has been dominated by hydropower development since the early 1980s. The country's hydropower expansion must necessarily be viewed with concern, given its magnitude, its environmental consequences, the debt it has contributed to the public purse, and the limited contribution that hydropower makes to Lao society and economy. The central argument is that the impetus for Laos's hydropower buildout is its construction and the opportunities that it creates for corruption. Corruption is central to Lao hydropolitics, and in turn, central to the conduct and practice of the Lao state.

State, party, and patronage

The countries of the Mekong Region "feed off the shared imperative of prioritising national modernisation and economic growth above all else" (Yeophantong, 2016: 290). In Laos, "mass infrastructure projects are seen as a way of both quick and visible development" (Wilcox, 2019: 169). The construction industry has become pivotal to this narrative "as a central nexus between private capital and public power" (Elinoff, 2017: 589). Large-scale infrastructure is cast as inherently good—so much so that it is beyond reproach. These "enchantments of infrastructure" (Harvey and Knox, 2012) underpin the country's messaging, proclaiming Laos as 'the battery of Southeast Asia' and keeping alive the promise of graduation from Least Developed Country status.

In the Mekong Region, all countries are to varying degrees authoritarian, which tends to support this type of neoliberal development imperative. There is a mutually reinforcing relationship between corruption and authoritarianism. Laos is no exception. "The nature of corruption is tied to the operation of the one-party political system in Laos. Following from that, it is unlikely to be solvable in its entirety unless we see some sort of political change accompanying it as well" (Hutt, 2019). Within the country, "major political and economic interests are virtually indistinguishable" (Simpson, 2007: 540). "Corruption", writes High (2012: 104), "is seen as the system itself, as part of the moral make-up of this reified entity, the Lao state … Corruption is one of the expectations of office now". The 1975 revolution overthrew the monarchy and established the LPRP as the country's sole political party. The nucleus of the LPRP is its Politburo which, in turn, is ensconced in its Central Committee. The Politburo assumes the role and function of the Central

Committee in between plenary meetings. Because the latter are infrequent, the Politburo is *de facto* Laos's highest decision-making authority. Its membership is drawn from the Central Committee, and the current (11th) Politburo comprises 13 members.

In contemporary Laos, association with the revolution justifies holding public office. Office holders without established revolutionary credentials are rare in the upper echelons of the LPRP. Responsibility for ruling has fallen on the children of revolutionaries. Thus, the ministers of finance and foreign affairs, and the president of the national assembly, are offspring of Kaysone Phomvihane. Similarly, Prime Minister Sonexay Siphandone is the son of revolutionary leader Khamtay Siphandone, while his sister is President of the Supreme People's Court, one brother is governor of Luang Namtha Province, and another was, until recently, mayor of Vientiane. These nepotistic groupings at party level extend to provinces from which much of their power is derived. Before 2006, governors were automatically members of the Central Committee. While this is no longer the case, governors continue to exercise considerable autonomy over business dealings within their provinces. This has prompted Croissant (2022) to argue that Laos is one of the most decentralised authoritarian regimes in the world. Internal Central Committee and Politburo politics reflects the competition between members and their provincial interests—most basically, between the northern and southern 'houses' of party royalty.

The southern house—currently the most powerful—is dominated by the Siphandone family. Khamtay Siphandone served as both Prime Minister and LPRP General Secretary. In the south, Khamtay amassed a fortune in construction business that enjoyed monopolistic control sanctioned by the government (Soukamneuth, 2006). At his death in 2025, aged 101, he was considered Laos's wealthiest man with an estimated net worth of USD1 billion (Bultmann, 2023). "Siphandone and his powerful family network are … notorious for acting primarily for personal gain" (Bertelsmann Stiftung, 2024: 28). The current president, Thongloun Sisoulith, is a northerner who married into the powerful Vongvichit family, reinforcing modest revolutionary credentials. "The socialist 'power elite' in the Central Committee wields power in the state and merges economic power with political domination", writes Bultmann, "In Laos, being powerful means being rich, and vice versa … Everyone rising in influence and economic power as an entrepreneur must be part of the patrimonial exchange with the established power elite and be a member of the state party" (2023: 24).

Lao corruption is widely acknowledged. In 2024, it scored 33 on Transparency International's Corruption Perception Index (where 0 indicates highly corrupt) and was ranked 114th of 180 countries (Transparency International, 2025). In 2024, the State Inspection and Anti-Corruption Authority (SIAA) claimed that the country had lost USD 32.1 million to corruption, recovering approximately USD 3.6 million

(Sitthixay, 2024). Notable cases included one in Luang Namtha Province, where 13 public officials, including a former provincial vice-governor, were prosecuted.

Laos's hydropower economy

Laos has a theoretical hydropower potential of 26.5 Gigawatts (GW), of which 18 GW is considered technically exploitable (IHA, 2016). Its first HPP was the 1 MW Nam Dong in the north, completed in 1970. Since then, installed hydropower capacity has grown to some 9,390 MW in 2023 (MERFI, 2025), comprising 75 dams (Fig. 5.2). An additional 3,177.5 MW are under construction. In 2023, Laos produced 52,295 TWh of electricity, of which 76 per cent came from hydropower and 23 per cent from coal (EIA, 2025). About three-quarters of Lao electricity is exported, primarily to Thailand, Cambodia, and Singapore. Electricity exports were worth USD 2.15 billion in 2023, with exports to Thailand worth USD 1.95 billion (OEC, 2024). Power production contributed about 12 per cent to national GDP in 2024, while electricity exports comprised almost 24 per cent of total export values (BOL, 2025).

Four characteristics of Laos's hydropower economy are fundamental to subsequent arguments. First, oversupply: Laos's electricity overproduction has been widely noted (ADB, 2019; Barney and Souksakoun, 2021). Oversupply is mainly a problem during the rainy season when water supply is plentiful. In 2021, Laos had an estimated surplus of 8,100 GWh, projected to grow to 16,900 GWh by 2025 (World Bank, 2023). Despite the staggering buildout, this was not accompanied by similar investments in transmission infrastructure. Laos must import electricity from Thailand during dry seasons.

Second, exaggerated demand projections: Power Development Plans (PDPs) have high political potency. High rates of predicted demand justify continued investments. In 2016, 5,062 GWh was consumed; projections had anticipated 6,131 GWh (JICA, 2020). The current PDP estimated consumption of 15,464 GWh in 2024—almost 45 per cent over the 2021 figure and a compound annual growth rate of 6.8 per cent. High overestimation appears chronic in Laos (Souksakoun, 2022).

Third, high externalities: environmental impacts include effects on hydrology (Hecht et al., 2019), sediments (Pukinskis, 2013), fisheries (Pukinskis and Geheb, 2012), aquatic biodiversity, climate (Räsänen et al., 2018), and forests. Hydropower usually requires resettlement and compensation of affected people. In Laos, these costs are largely transferred to society. Thailand effectively 'exports externalities' by transferring costs of power consumption to Laos (Simpson, 2007; Marks and Zhang, 2019).

Fourth, limited benefits to society: while power production contributes significantly to GDP, GDP growth contributes only modestly to poverty alleviation (World Bank, 2017). Although poverty has been alleviated over two decades, this is mainly

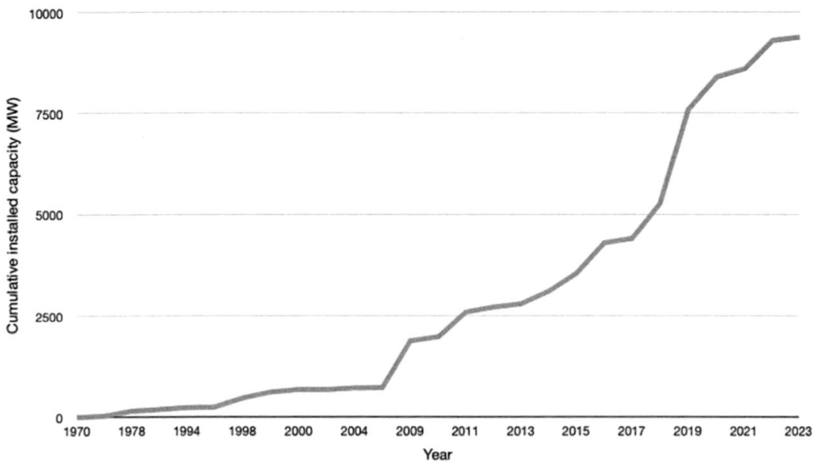

Figure 5.2: Growth in installed hydropower capacity in Laos (1970-2023) (generated from data in MERFI, 2025).

because of agricultural commercialisation (World Bank, 2017). Laos's resource-driven approach "has created few jobs, while leading to significant environmental and macroeconomic pressures" (World Bank, 2021). Income inequality remains essentially unchanged since 1990, with 47.5 per cent of national income held by the top 10 per cent and 14.2 per cent by the bottom 50 per cent (WID, 2024).

The justifications for continued excessive emphasis on hydropower are not presently—and have not been for some time—strong. This suggests that motivation lies beyond assertions about national development and public revenue. One strong candidate is that construction itself is attractive. "Laos's hydropower plans appear to have become untethered from any solid economic rationale, driven overwhelmingly by the domestic interests that stand to benefit from dam *construction*" (Strangio, 2021, emphasis added).

Early engagements between the Lao state and multilateral development banks also introduced Laos to the linguistics of 'sustainable hydropower'. The 1,070 MW Nam Theun 2 (NT2), supported by the World Bank and ADB, was intended to demonstrate that sustainable hydropower was no oxymoron—a "kinder, gentler dam" (Kent, 2007). The institutional reforms accompanying NT2 created most of Laos's environmental legislation (Goldman, 2001). Yet whenever the GOL wishes to display its sustainability credentials, it invokes NT2 while deeming the replication of its standards unfeasible elsewhere. As then-Vice-Minister Viraphonh Viravong stated: "To use it as a standard, it's not possible. We can use it as a good example, a good guideline, but not as a standard. All the developers say that it is not possible to use Nam Theun 2 as a standard" (quoted in Jusi, 2011: 256).

Domestic hydrocorruption

Initial experience with THPC and subsequent early BOT ventures laid the groundwork for the Lao hydrocorruption model. In the initial capital burn that hydropower construction requires, significant opportunities arise for corruption. In 2012, Matthews estimated that 5–20 per cent of hydropower costs are lost to corruption depending on the project size. Companies frequently collude through direct bribery or by inflating input costs (Souksakoun, 2022).

The perceived success of ventures such as THPC and NT2 initiated a development trajectory that would dominate Lao narratives for four decades. "When the LPRP came to power … it embraced hydropower as a key ideological symbol of its own industrial ambitions, emblazoning the national emblem with the dam" (Creak and Barney, 2022: 15–16). 'Capturing' the hydropower sector suggested two strategies: developing local capabilities through which LPRP interests could be funnelled, and forming relationships with foreign firms keen on the construction opportunities that Lao rivers offer. These were never mutually exclusive.

Initially, the locus of Lao power investments was Électricité du Lao (EdL). Founded in 1959, this state-owned enterprise was ostensibly overseen by the Ministry of Energy and Mines (MEM). As an SOE, EdL is responsible for its direct debts—not the GOL. This meant that EdL became a useful vessel for channelling poorly conceived loans. "[N]o transparency or checks or balances exist in EdL's bidding process, resulting in higher-than usual project costs and unmanageable levels of interest and debt", said sources quoted by Souksavanh (2021). As a consequence, EdL has been a significant driver of debt accumulation.

In 2023, Lao external debt stood at an estimated USD 17 billion, some 112 per cent of GDP. Much was on-lent to EdL, whose total debt load was USD 5.4 billion, including USD 1.8 billion in government-guaranteed debt. A further USD 1 billion in non-guaranteed debt was held by EdL-Generation (Barney et al., 2025). The energy sector therefore holds almost 38 per cent of sovereign debt. Should EdL default, liability shifts to the GOL. The current strategy appears to ensure that EdL does not default.

EdL spawned two major subsidiaries: EdL-Generation (EdL-GEN), responsible for power generation; and EdL-Transmission (EdL-T), responsible for the grid. EdL-GEN owns ten HPPs with a combined installed capacity of 526 MW, and shareholdings in 16 additional HPPs totalling 4,333 MW (EdL-GEN, 2024). Dams producing for the domestic market feed into the Lao grid under 'take-or-pay' deals. Here, EdL guarantees the purchase of certain amounts from producers at set rates, even if surplus to demand. These commitments have contributed significantly to EdL's indebtedness (Barney and Souksakoun, 2021), and Souksakoun (2022: 111–112) suggests EdL may have received political pressure to accept them "due to their individual connection to hydropower projects and their personal economic interests". If the focus is on constructing dams rather than the subsequent revenue, financial performance after commissioning is a relatively redundant consideration. In July 2025, EdL announced that it would renegotiate take-or-pay rates due to "growing public anger over rising power bills" (Thongsavanh, 2025).

Besides EdL, several private firms serve as commercial fronts for LPRP interests—a "step in the legitimisation" of corrupt practices, as one source described it. Chaleun Sekong Group (CS Group), with substantial hydropower interests through Chaleun Sekong Energy (CSE), is thought to fall under Prime Minister Siphandone's patronage. CSE's asset base of seven wholly owned dams and three shareholdings is distributed mainly in southern provinces where Siphandone's interests lie. Phongsupthavy Group, under President Thongloun Sisoulith's patronage, was originally focused on road and bridge construction. Unable to pay for completed work, the GOL awarded hydropower concessions instead. Phongsupthavy offers services in construction, agriculture, and bitcoin mining.

Until recently, Phongsupthavy held 24 per cent of shares in EdL-GEN, giving its owner a board seat. It appears to have had an 'inside track' on contracts to supply the Vietnamese market via Mekong Investment Holdings Co., in which Phongsupthavy is a shareholder (Wikipedia, n.d.). Through this firm, Phongsupthavy maintained close relations with former Vietnamese Prime Minister Nguyễn Xuân Phúc, who resigned in 2023 over corruption allegations. This relationship allegedly obliged Electricity Vietnam (EVN) to sign numerous PPAs with Phongsupthavy, which promised to construct dams. Very few materialised; Phongsupthavy would partially commence construction before selling concessions to other firms like CSE. "EVN hates Phongsupthavy because they never finish anything", one source observed.

In January 2015, citizens were 'raising eyebrows' at President Choummaly Sayasone's new house in Ban That Luang, which came with a USD 650,000 garden. The contract was issued to Phonesack Group, 'which has a good relationship with Choummaly's family' (RFA, 2015). Phonesack's owner is closely related to the powerful Pholsena family and is said to be 'untouchable' (Boh, 2013). The Phonesack Group is better known for alleged illegal logging activities (EIA, 2012). During 2006-2007, the national timber quota was raised to allow clearance of NT2's reservoir; Phonesack was involved (EIA, 2014) and may have seen hydropower as a means to circumvent logging bans—a strategy gaining popularity across the region.

The Xe Pian–Xe Nam Noi HPP is located on the Bolaven Plateau in Attapeu Province. In July 2018, following heavy rains, an auxiliary dam collapsed, sending an estimated 350 million cubic metres of water downstream (Latrubesse et al., 2020). Nineteen villages were inundated, 71 people killed, and 14,400 displaced (RFA, 2023). Rescue teams from across the region arrived, and millions in financial aid flowed from sympathetic countries and the Lao public. Dam owners claimed to have paid USD 91.2 million in compensation, 71 per cent to the Lao state (International Rivers,

2023). In January 2024, it was revealed that USD 4.2 million intended for survivors had 'gone missing'. A provincial employee commented: "The money must have been used inappropriately" (RFA, 2024).

Following the disaster, the GOL called for a safety audit. A nationwide inspection of dams with more than 15 MW capacity was carried out by an international team. Of 56 dams, 21 had to be excluded because 'insufficient basic data were missing to give a critical assessment'. Of the remaining 35, findings included (MEM, 2020): seismic hazards had been inadequately addressed in 46 per cent; in 55 per cent, operators rarely opened bottom outlets for safety drawdown; in 65 per cent, slope inspection was insufficient; in 29 per cent, cracks or deteriorated joints were noticed; and for 98 per cent, emergency action plans were inadequate or absent. It is difficult to imagine how, absent corrupt practice or gross negligence, such transgressions can be explained.

In March 2025, a former EdL deputy director and four contractors faced prosecution for corruption involving the Nam Hinboun Project, owned by EdL and financed with a USD 90 million loan from a Thai bank. "Authorities discovered that the construction was only 76 percent complete, despite the project owner having already paid the contractor 98.64 percent of the total contract value" (Chanlivong, 2025). EdL-owned projects appear particularly susceptible to corrupt practices.

Transnational hydrocorruption

In 1991, Somsavat Lengsavath joined the LPRP Central Committee. Born in Luang Prabang to Chinese and Lao parents, Somsavat was a protégé of Phoumi Vongvichit, a founding LPRP member and key Politburo member (Stuart-Fox, 2013). Due to his parentage, he was the only Mandarin-speaking Politburo member. Following China's attempted invasion of Vietnam in 1979, Laos had largely withdrawn connections with China. But Somsavat changed this, persuading the Politburo to open relations with their northern neighbour amid the Asian financial crisis and Laos's economic difficulties. He "played a significant role in drawing Chinese investors into Laos … and developed close personal relationships with counterparts in China" (Aroonpipat, 2018: 58).

Somsavat's network included Zhao Wei, who holds the concession for the Golden Triangle Special Economic Zone in Bokeo Province, where his Kings Romans casino operates. Zhao and the Kings Romans Group have been sanctioned by the US as a 'transnational criminal organization' for alleged involvement in "drug trafficking, human trafficking, money laundering, bribery, and wildlife trafficking" (Thorley, 2024: 17). The casino derives power from the Chinese-constructed 168 MW Nam Tha 1 dam just east of the SEZ.

As relations thawed, Laos established a Cooperation Commission for Sino-Lao aid coordination. Somsavat became president and ensured that authorisations for Chinese funding bypassed the Ministry of Planning and Investment (MPI), extending directly from the Commission to himself (Aroonpipat, 2018). Sources claim that Somsavat exploited this position to the fullest degree, receiving various Chinese inducements to support their interests within the Politburo and to authorise investments.

Somsavat's initiatives included the Kunming-Vientiane railway, which he "almost single-handedly ushered through the Lao bureaucracy" (Doig, 2019). He also secured a USD 100 million loan from China Development Bank for the 2009 Southeast Asian Games stadium. In return, the bank would receive 1,600 hectares around That Luang Marsh for development into a satellite city (Stuart-Fox, 2009; Chen, 2020). There were 17 villages and nearly 38,000 people located around the marsh requiring resettlement.

> As is common with development projects in Laos, the developer handed over a sum of money to the government that would be used for compensation, and the government was responsible for doling it out. There were of course a range of inconsistencies ... that led landowners to increasingly believe that government officials were lowering compensation rates and skimming the rest off the top. There wasn't really any transparency on how compensation rates were determined (Kenney-Lazar and Chen, 2022).

At the 10th LPRP Congress in 2016, Somsavat was retired—speculatively because he had swung the pendulum too far toward China.

The 480 MW Nam Ngum 3 was another Somsavat initiative. Initially, the ADB had been keen to finance it, "but their safeguard standards rendered them uncompetitive with the GoL" (Souksakoun, 2022: 28). China's Exim Bank provided a USD 1.29 billion loan. Somsavat allegedly received USD 50 million for negotiating the loan, drawn from the initial tranche. Payments were suspended and construction halted—not because of the payment to Somsavat, but because EdL transferred shares to CS Group contrary to contractual obligations. CS Group entered a joint venture with EGAT International, restarting construction. In late 2024, EdL reportedly resumed 100 per cent ownership but still requires USD 300 million for completion. Remarkably, it is rumoured to have obtained USD 100 million from the Kings Romans Casino Group, channelled through a Vientiane-based Chinese firm. Allegedly, Kings Romans's interest is money laundering.

In June 2012, religious ceremonies were held for groundbreaking of the 1,285 MW Xayaburi dam. Then-LPRP Secretary General Choummaly Sayasone presided. The developer, CH Karnchang, is Thailand's second-largest construction firm whose "political connections run deep. Some 70 percent of its revenues are derived from public sector and military construction projects" (Wingfield, 2002: 293). CH Karnchang's first foray into Lao hydropower was the 615 MW Nam Ngum 2, a concession that had been languishing. Significant corruption allegedly occurred around this concession. The Xayaburi was its second venture, Laos's first Mekong mainstream dam.

CH Karnchang allegedly heavily influenced Laos's decision to proceed despite significant environmental concerns and before regional consultations in the Mekong River Commission were complete, thereby endangering Laos's relations with neighbours (Geheb et al., 2015). Choummaly allegedly received substantial payments:

> He gets big money (from Karnchang). He cannot pay it back as it is all spent. The dam will go ahead. He has a full time staff person whose job it is to convey his wishes to CH Karnchang. If his daughter wants a laptop, it is there within days. The reason this will go ahead despite the opposition, is that the very top is getting the money (Boh, 2012).

In the south, Malaysian investors allegedly issued significant inducements to then-Governor of Champasak Province, Sonexay Siphandone, for the Don Sahong mainstream dam. In a 2010 meeting with provincial officials, the governor allegedly stated that the dam was good for the country and "anyone who opposed it was an enemy of the people". The 'behind-the-scenes' project manager was Sonexay's son-in-law, and the Siphandone family was "expected to gain significant benefits" if the project proceeded (Baird, 2011). The relevance of the project to the Siphandone family was said to have been so great that Sonexay avoided a transfer to a ministerial post "possibly because he may be in a better position to protect his family's interests in the Don Sahong Dam by remaining in power in Champasak" (Baird, 2011: 44).

Other significant transnational arrangements warrant mention. The Electricity Generating Authority of Thailand (EGAT) rotates senior staff through subsidiaries including the RATCH Group, EGAT International, and the EGCO Group. Between them, they hold shares in seven major Lao hydropower projects, all selling power to EGAT. Having subsidiaries sell to parent firms creates perverse incentives to negotiate prices down—with impacts on Lao government revenues and clear conflicts of interest (MEE-Net, 2013). This arrangement "incentivizes collusion and the prospects of agreed cost structures rather than competition with incentives to lower costs" (Jarvis, 2010: 184). RATCH's shareholding in EdL-GEN also positions it to influence prices of electricity sold by EdL-GEN to the Thai grid.

Equally intriguing is the relationship between Chinese developers of three planned mainstream dams and Thailand's Gulf Energy Development PLC (GED). Chinese developers seeking Thai market access require Thai partners for EGAT power purchase agreements. They appear to have settled on GED as their single entry point. GED also holds shares in CH Karnchang's current venture, the 1,460 MW Luang Prabang dam. GED's CEO, Sarath Ratanavadi, Thailand's wealthiest man, has rumoured connections to Thaksin Shinawatra, under whose watch GED obtained construction contracts for 5,300 MW of gas turbines in a deal considered highly irregular (Macan-Markar, 2021).

Discussion and conclusions

In the examples above, I have paid attention to those hydropower cases for which evidence is most available. Absence of evidence is not evidence of absence; many other projects and corrupt partnerships frequently surface in conversations with sources. A Politburo already accustomed to patrimony and corrupt natural resource exploitation—particularly through logging, which dried up due to international attention—sought to capture the hydropower sector. The perceived success of early ventures initiated a development trajectory that would dominate Lao narratives for four decades.

This diffuse institutional landscape—EdL, EdL-GEN, the Lao Holding State Enterprise, MPI—may reflect state reactions to needs as they arise but may also provide opportunities for capture by individual ministries and the Politburo. Coupled with deliberate secrecy, this creates the ideal conditions for hydrocorruption. No one really knows which company will be awarded which concession or which construction firm has swayed which Politburo member's opinion. The MEM project list is not public; neither is the Power Development Plan. There are dozens of dams for which environmental assessments are unavailable, financing unknown, design reviews inaccessible, and authorisation processes obscure.

Hydropower also intersects with state territorial agendas. Laos recognises 50 ethno-linguistic groups; approximately 47 per cent of the population is ethnically unrelated to the dominant Lao. Ethnic minorities predominantly occupy highlands—precisely where most hydropower potential lies. The World Bank (2010) estimated hydropower would resettle 100,000-200,000 people; the then-Minister of Energy and Mines estimated 60,000 people from 12,000 families had been resettled by 2019 (Vientiane Times, 2019). Resettlement budgets provide further corruption opportunities when hydropower firms delegate compensation distribution to district or provincial authorities—and funds disappear.

Recent scholarship suggests that contemporary resettlement dynamics have evolved. Ponce (2022, 2023, 2025), drawing on ethnographic research at a Chinese hydropower resettlement site in Bokeo Province, demonstrates that Lao hydropower politics have moved beyond concerns with sedentarisation and cultural change toward processes of urbanisation. Resettlement now often involves constructing new townships with modern infrastructure—electricity, roads, housing—offering similar rent-seeking opportunities to dam construction itself. Building resettlement towns, Ponce's work suggests, provides parallel incentives for political elites promoting large construction projects. The territorial logic of hydropower thus reinforces its corruption logic: both the dam and the displacement which it necessitates generate construction opportunities.

The hydrocorruption framework which I have outlined has several analytical advantages. First, it foregrounds corruption as central rather than incidental to hydropower development: not a distortion of an otherwise rational development process but a structuring logic. Second, it draws attention to how the particular characteristics of hydropower—its capital intensity, complexity, opacity, and path dependencies—make it especially amenable to capture. Third, it connects domestic and transnational dimensions, showing how Lao hydrocorruption operates through networks linking the Politburo, provincial authorities, domestic firms, Thai and Chinese corporations, and international financiers. Fourth, it helps explain patterns otherwise difficult to account for: persistent dam-building despite chronic oversupply, systematic flouting of safety standards, and opacity surrounding concessions and contracts.

This analysis suggests that hydropower development in Laos is under-monitored, uncoordinated, below safety standards, and makes remarkably little contribution to its socio-economy. It has, however, benefited many Lao politicians, Politburo members, and connected Lao and international firms. The 'enchantments of infrastructure' (Harvey and Knox, 2012) underpin messaging proclaiming Laos as 'the battery of Southeast Asia'. The hydrocorruption framework suggests that this narrative functions as legitimation for what is, fundamentally, a rent-seeking project.

Whether and how this might change remains uncertain. EdL's mounting debt crisis, a PowerChina lawsuit for USD 555 million in unpaid dues (Reuters, 2025), and recent prosecutions of EdL officials may indicate pressures within the system. But the fundamental structure—single-party rule, patronage networks, opacity, and the continuing allure of construction—remains intact.

References

ADB (Asian Development Bank). 1997. Hydropower project to increase Lao PDR's GDP by 7 percent. ADB Review. Mandaluyong City: ADB.

ADB (Asian Development Bank). 2019. Lao People's Democratic Republic energy sector assessment, strategy, and road map. Mandaluyong City: ADB, https://www.adb.org/sites/default/files/institutional-document/547396/lao-pdr-energy-assessment-2019.pdf (accessed 13 July 2025).

Ansar, A., Flyvberg, B., Budzier, A. and Lunn, D. 2014. Should we build more large dams? The actual costs of hydropower megaproject development. *Energy Policy* 69(2014): 43–56, https://doi.org/10.1016/j.enpol.2013.10.069.

Aroonpipat, S. 2018. Governing aid from China through embedded informality: institutional response to Chinese development aid in Laos. *China Information* 32(1): 46–68, https://doi.org/10.1177/0920203X17730330.

Baird, I.G. 2011. Laos. In Lee, P.O. and Montesano, M.J. (eds.), *Regional outlook: Southeast Asia, 2011-2012*, 43–47. Singapore: Institute of Southeast Asian Studies.

Barney, K. and Souksakoun, K. 2021. Credit crunch: Chinese infrastructure lending and Lao sovereign debt. *Asia & the Pacific Policy Studies* 8(1): 94–113, https://doi.org/10.1002/app5.318.

Barney, K., Rajah, R. and Cooray, M. 2025. Trapped in debt: China's role in Laos' economic crisis. Sydney: Lowy Institute, https://cdn.sanity.io/files/5dcouzig/production/196200e4f821953fb0e71a339f2c375d39e1d5d1.pdf (accessed 13 July 2025).

Bertelsmann Stiftung. 2024. BTI 2024 Country Report – Laos. Gütersloh: Bertelsmann Stiftung, https://bti-project.org/fileadmin/api/content/en/downloads/reports/country_report_2024_LAO.pdf (accessed 13 July 2025).

Boh, M. 2012. Wet dreaming. *On Line Opinion*, 1 August 2012, https://www.onlineopinion.com.au/view.asp?article=13933&page=1 (accessed 13 July 2025).

Boh, M. 2013. A dam too far in Laos. *Asia Times Online* 12 April 2013, https://riverresourcehub.org/resources/a-dam-too-far-in-laos-7925/ (accessed 12 July 2021).

BOL (Bank of the Lao PDR). 2025. Annual Economic Report 2024. Vientiane: BOL, https://bol.gov.la/en/fileupload/20-06-2025_1750414260.pdf (accessed 13 July 2025).

Bultmann, D. 2023. Conflict and elite formation in Thailand, Laos and Cambodia. *TRaNS: Trans-Regional and National Studies of Southeast Asia* 11(1): 17–29, https://doi.org/10.1017/trn.2022.7.

Chanlivong, K. 2025. Former Electricité du Laos officials, contractors face prosecution over corruption. *The Laotian Times* 21 March 2025, https://laotiantimes.com/2025/03/21/former-electricite-du-laos-officials-contractors-face-prosecution-over-corruption/ (accessed 13 July 2025).

Chen, W. 2020a. The power of mirage: state, capital, and politics in the grounding of "Belt and Road" in Laos. PhD Thesis. Madison (WI): University of Wisconsin, https://asset.library.wisc.edu/1711.dl/DDNYGF2NP4CQB8W/R/file-8d0a5.pdf (accessed 10 July 2025).

Creak, S. and Barney, K. 2022. The role of 'resources' in regime durability in Laos. *Communist and Post-Communist Studies* 55(4): 35-58, https://doi.org/10.1525/cpcs.2022.1713051.

Croissant, A. 2022. *Comparative politics of Southeast Asia: an introduction to governments and political regimes*. Cham: Springer International Publishing, https://doi.org/10.1007/978-3-031-05114-2.

Doig, W. 2019. The Belt and Road Initiative is a corruption bonanza. *Foreign Policy* 15 January 2019, https://foreignpolicy.com/2019/01/15/the-belt-and-road-initiative-is-a-corruption-bonanza/ (accessed 21 November 2024).

EdL-GEN (Electricité du Lao-Generation). 2024. Our Business – Hydro Power. Vientiane: EdL-GEN, https://edlgen.com.la/ (accessed 2 July 2025).

EIA (Environmental Investigation Agency). 2012. Checkpoints: how powerful interest groups continue to undermine forest governance in Laos. London: EIA, https://eia-international.org/wp-content/uploads/EIA-Crossroads-report-FINAL-low1.pdf (accessed 11 May 2025).

EIA (Environmental Investigation Agency). 2014. Routes of extinction: the corruption and violence destroying Siamese rosewood in the Mekong. London: Environmental Investigation Agency, https://eia-international.org/wp-content/uploads/Routes-of-Extinction-FINAL-lo-res.pdf (accessed 13 July 2025).

EIA (Energy Information Administration). 2025. Laos. Washington D.C.: EIA, https://www.eia.gov/international/data/country/LAO/electricity/electricity-generation?pd=2&p=00000000000000000000000000000fvu00000000000000000000000000000000u&u=0&f=A&v=mapbubble&a=-&i=none&vo=value&t=C&g=none&l=249--124&s=315532800000&e=1672531200000 (accessed 10 July 2025).

Elinoff, E. 2017. Concrete and corruption: materialising power and politics in the Thai capital. *City* 21(5): 587–596, https://doi.org/10.1080/13604813.2017.1374778.

Geheb, K., West, N. and Matthews, N. 2015. The Invisible Dam: Hydropower and its narration in the Lao Peoples' Democratic Republic. In Matthews, N. and Geheb, K. (eds.), *Hydropower Development in the Mekong Region: political, socio-economic and environmental perspectives*, 101–126. London: Earthscan.

Goldman, M. 2001. Constructing an environmental state: eco-governmentality and other transnational practices of a "green" World Bank. *Social Problems* 48(4): 499–523, https://doi.org/10.1525/sp.2001.48.4.499.

Haas, L.J.M. 2008. Water for energy: corruption in the hydropower sector. In Transparency International (ed.), *Global corruption report 2008*, 85–86. Cambridge: Cambridge University Press.

Harvey, P. and Knox, H. 2012. The enchantments of infrastructure. *Mobilities* 7(4): 521–536, https://doi.org/10.1080/17450101.2012.718935.

Hecht, J.S., Lacombe, G., Arias, M.E., Dang, T.D. and Piman, T. 2019. Hydropower dams of the Mekong River basin: a review of their hydrological impacts. *Journal of Hydrology* 568(2019): 285–300, https://doi.org/10.1016/j.jhydrol.2018.10.045.

High, H. 2012. *Fields of desire: poverty and policy in Laos*. Singapore: NUS Press.

Hutt, D. 2019. The only way to end corruption in Laos. *The Diplomat* 10 July 2019, https://thediplomat.com/2019/07/the-only-way-to-end-corruption-in-laos/ (accessed 5 May 2025).

IHA (International Hydropower Association). 2016. 2016 Hydropower Status Report. London: International Hydropower Association, https://www.aler-renovaveis.org/contents/lerpublication/iha_2016_may_hydropower-status-report-2016.pdf (accessed 13 July 2025).

International Rivers. 2023. PRESS RELEASE – Joint Statement: Lao Dam Disaster – NGOs decry lack of effective remedy for survivors five years on. Berkely: International Rivers, https://www.internationalrivers.org/news/press-release-joint-statement-ngos-decry-lack-of-effective-remedy-for-survivors-five-years-on/ (accessed 13 July 2025).

Jarvis, D.S.L. 2010. Institutional processes and regulatory risk: a case study of the Thai energy sector. *Regulation & Governance* 2010(4): 175–202, https://doi.org/doi:10.1111/j.1748-5991.2010.01077.x.

JICA (Japan International Cooperation Agency). 2020. The study on Power Network System Master Plan in Lao People's Democratic Republic – Final Report. Tokyo: Japan International Cooperation Agency, https://openjicareport.jica.go.jp/pdf/12328027_01.pdf (accessed 10 July 2025).

Johnson, K. and Wongcha-Um, P. 2020. Amid hydropower boom, Laos streams ahead on latest Mekong dam. *Reuters* 7 February 2020, https://www.reuters.com/article/us-mekong-river-dam/amid-hydropower-boom-laos-streams-ahead-on-latest-mekong-dam-idUSKBN2010B8/?il=0 (accessed 13 July 2025).

Jusi, S. 2011. Challenges in developing sustainable hydropower in Lao PDR. *International Journal of Development Issues* 10(3): 251–267, https://doi.org/10.1108/14468951111165377.

Kenney-Lazar, M. and Chen, W. 2022. Dispossession and resistance in an urbanizing wetland of Vientiane, Laos. NUS Southeast Asian Studies Online Webinar Series, https://www.youtube.com/watch?v=0qlEL_i_2ng&t=60s (accessed 3 July 2025).

Kent, J. 2007. Laos dam raises ethical bar. *Newsweek* 11 August 2007, http://www.newsweek.com/laos-dam-raises-ethical-bar-99271 (accessed 15 July 2025).

Latrubesse, E.M., Park, E., Sieh, K., Dang, T., Lin, Y.N. and Yun, S.-H. 2020. Dam failure and a catastrophic flood in the Mekong basin (Bolaven Plateau), southern Laos, 2018. *Geomorphology* 362: 107221, https://doi.org/10.1016/j.geomorph.2020.107221.

Macan-Markar, M. 2021. Thailand's billionaire boom: the rise of Sarath Ratanavadi. *NikkeiAsia* 21 November 2021, https://asia.nikkei.com/Spotlight/The-Big-Story/Thailand-s-billionaire-boom-the-rise-of-Sarath-Ratanavadi (accessed 14 July 2025).

Marks, D. and Zhang, J. 2019. Circuits of power: environmental injustice from Bangkok's shopping malls to Laos' hydropower dams. *Asia Pacific Viewpoint* 60(3): 296–309, https://doi.org/10.1111/apv.12242.

Matthews, N. 2012. Water grabbing in the Mekong basin: an analysis of the winners and losers of Thailand's hydropower development in Lao PDR. *Water Alternatives* 5(2): 392-411, http://www.water-alternatives.org/index.php/volume5/v5issue2/176-a5-2-12/file.

MEE-NET (Mekong Energy and Ecology Network). 2013. Following the money trail of Mekong energy industry. Bangkok: MEE-NET, https://laofab.org/document/download/2398 (accessed 8 January 2026).

MEM (Ministry of Energy and Mines). 2020. Review of the nationwide emergency dams safety inspection: safety assessment review – synthesis report. Confidential report. Vientiane: MEM.

MERFI (Mekong Region Futures Institute). 2025. Dataset on the Dams of the Greater Mekong. Bangkok: MERFI, https://www.merfi.org/mekong-region-dams-database (accessed 18 April 2024).

Mutebi, A.M. 2008. Explaining the failure of Thailand's anti-corruption Regime. *Development and Change* 39(1): 147–171, https://doi.org/10.1111/j.1467-7660.2008.00472.x.

OEC (Observatory of Economic Complexity). 2024. Electricity in Laos, https://oec.world/en/profile/bilateral-product/electricity/reporter/lao (accessed 13 July 2025).

Ponce, F.S.J. 2022. 'Eating with the people': how a Chinese hydropower project changed food experiences in a Lao community. *Social Anthropology* 30(1): 1–23, https://doi.org/10.3167/saas.2022.300102.

Ponce, F.S.J. 2023. Hydroelectric development in 'China's backyard'? *Asian Anthropology* 22(4): 298–302, https://doi.org/10.1080/1683478X.2023.2254129.

Ponce, F.S.J. 2025. Beyond debt-trap narratives: how the Lao State manoeuvres the hybrid governance of a Chinese hydropower project. *Advances in Southeast Asian Studies* 18(1): 5–29, https://doi.org/10.14764/10.ASEAS-0118.

Pukinskis, I. 2013. Mekong sediment basics. *State of Knowledge* 2. Vientiane: CGIAR Challenge Program on Water and Food, https://cgspace.cgiar.org/server/api/core/bitstreams/1a31b0bb-073b-4009-bac1-57fbedd6c2b7/content (accessed 24 June 2026).

Pukinskis, I. and Geheb, K. 2012. The impacts of dams on the fisheries of the Mekong. *State of Knowledge* 1, Vientiane: CGIAR Challenge Program on Water and Food, https://cgspace.cgiar.org/rest/bitstreams/24817/retrieve (accessed 13 July 2025).

Räsänen, T.A., Varis, O., Scherer, L. and Kummu, M. 2018. Greenhouse gas emissions of hydropower in the Mekong River Basin. *Environmental Research Letters* 13(3): 034030, https://doi.org/10.1088/1748-9326/aaa817.

Reuters. 2025. Exclusive: PowerChina unit sues Laos utility for $555 million. *Reuters* 5 March 2025, https://www.reuters.com/business/energy/powerchina-unit-sues-laos-utility-claims-555-million-unpaid-dues-2025-03-05/ (accessed 12 July 2025).

RFA (Radio Free Asia). 2015. Luxury Home Building by Lao Leaders Raises Eyebrows Among Citizens. *Radio Free Asia* 23 January 2015, https://www.rfa.org/english/news/laos/politicians-build-luxury-homes-01232015162351.html (accessed 12 July 2025).

RFA (Radio Free Asia). 2023. Five years on, compensation yet to arrive for Lao dam collapse survivors. *Radio Free Asia* 21 July 2023, https://www.rfa.org/english/news/laos/dam-07212023101019.html (accessed 13 July 2025).

RFA (Radio Free Asia). 2024. Laos governor: $4.2 million for dam collapse survivors has gone missing. *Radio Free Asia* 17 January 2024, https://www.rfa.org/english/news/laos/missing-funds-dam-collapse-01172024125553.html (accessed 15 July 2025).

Robbins, P. 2000. The rotten institution: corruption in natural resource management. *Political Geography* 19(4): 423–443, https://doi.org/10.1016/S0962-6298(99)00087-6.

Simpson, A. 2007. The environment – energy security nexus: critical analysis of an energy 'love triangle' in Southeast Asia. *Third World Quarterly* 28(3): 539–554, https://doi.org/10.1080/01436590701192710.

Sitthixay, P. 2024. Fighting corruption: Laos recovers millions in 2024. *The Laotian Times* 10 December 2024, https://laotiantimes.com/2024/12/10/fighting-corruption-laos-recovers-millions-in-2024/ (accessed 15 January 2025).

Sopera, D. 2022. Corruption and water governance in the Mekong River Basin. *U4 Issue* 2022:12. Bergen: U4 Anti-Corruption Resource Centre, Chr. Michelsen Institute, https://www.u4.no/publications/corruption-and-water-governance-in-the-mekong-river-basin.pdf (accessed 7 January 2026).

Soukamneuth, B.J. 2006. The political economy of transition in Laos: from peripheral socialism to the margins of global capital. PhD Thesis. Ithaca (NY): Cornell University, https://ecommons.cornell.edu/server/api/core/bitstreams/63ee0d05-b228-4e1c-a7c3-b7a90dddc8b6/content (accessed 13 July 2025).

Souksakoun, K. 2022. Towards sustainable hydropower: policy implementation and livelihood transformation in the Sekong Basin, Laos. PhD Thesis. Canberra: Australian National University, https://openresearch-repository.anu.edu.au/bitstream/1885/270288/1/Final%20thesis%20for%20submission_Kanya%20Souksakoun_2022.pdf (accessed 13 July 2025).

Souksavanh, O. 2021. Lao power company deep in debt, but managers live well. *Radio Free Asia* 10 May 2021, https://www.rfa.org/english/news/laos/debt-05102021112817.html (accessed 21 January 2025).

Stålgren, P. 2006. Corruption in the water sector: causes, consequences and potential reform. *Swedish Water House Policy Brief* 4. Stockholm: Swedish Water House, Swedish International Water Institute, https://siwi.org/wp-content/uploads/2015/03/20060906171639corruptionpb_low-3.pdf (accessed 13 July 2025).

Strangio, S. 2021. "Not sufficient": Thailand rejects report on Lao hydropower dam. *The Diplomat* https://thediplomat.com/2021/01/not-sufficient-thailand-rejects-report-on-lao-hydropower-dam/ (accessed 5 July 2025).

Stuart-Fox, M. 2013. *Historical dictionary of Laos*. Lanham: Scarecrow Press.

Thongsavanh, S. 2025. Lao government to renegotiate hydropower deals amid public outcry over electricity prices. *The Laotian Times* 18 July 2025, https://laotiantimes.com/2025/07/18/lao-government-to-renegotiate-hydropower-deals-amid-public-outcry-over-electricity-prices/ (accessed 23 July 2025).

Thorley, M. 2024. A changing landscape: China's new model of global governance and its impact on the fight against organized crime. Geneva: The Global Initiative Against Transnational Organized Crime, https://globalinitiative.net/wp-content/uploads/2024/05/Martin-Thorley-A-changing-landscape-Chinas-new-model-of-global-governance-and-its-impact-on-organized-crime-GI-TOC-May-2024.pdf (accessed 13 July 2025).

Transparency International. 2008. *Global corruption report 2008*. Cambridge: Cambridge University Press.

Transparency International. 2025. Corruption Perceptions Index 2024, https://www.transparency.org/en/cpi/2023/index/lao (accessed 10 July 2025).

Vientiane Times. 2019. 12,000 Lao families relocated for hydropower projects. *Vientiane Times* 29 March 2019, https://www.facebook.com/vientianetimesonline/photos/60000-relocated-for-hydro-power-projects-across-countryabout-us500-million-has-b/2305199132880860/ (accessed 8 January 2026).

Vientiane Times. 2023. Praise for Theun-Hinboun Power Company on 25th Anniversary. *Vientiane Times* 3 April 2023, https://www.vientianetimes.org.la/sub-new/Previous_65_y23/freeContent/FreeConten65_Praise_y23.php (accessed 13 July 2025).

WCD (World Commission on Dams). 2000. *Dams and development – a new framework for decision-making*. The Report of the World Commission on Dams. London: Earthscan.

WID (World Inequality Database). 2024. Income inequality, Lao PDR, 1980-2022. World Inequality Database, https://wid.world/country/lao-pdr/ (accessed 28 February 2025).

Wikipedia. n.d. Phongsavath Senaphuan. Wikipedia Talencyclopedia (Vietnamese), https://vi.wikipedia.org/wiki/Th%C3%Aonh_vi%C3%AAn:Talencyclopedia/Phongsavath_Senaphuan (accessed 20 November 2024).

Wilcox, P. 2019. *Heritage and the making of political legitimacy in Laos.* Amsterdam: Amsterdam University Press.

Williams, A. and Le Billon, P. 2017. Introduction. In Williams, A. and Le Billon, P. (eds.), *Corruption, natural resources and development: from resource curse to political ecology*, 1–13. Cheltenham, UK: Edward Elgar Publishing.

Wingfield, T. 2002. Democratization and economic crisis in Thailand: political business and the changing dynamic of the Thai state. In Gomez, T.E. (ed.), *Political Business in East Asia*, 250–300. London: Routledge.

World Bank. 2010. The socio-geography of mining and hydro in Lao PDR: analysis combining GIS information with socioeconomic data. Technical Note to the Lao PDR Development Report 2010: Natural Resource Management for Sustainable Development. Washington D.C.: World Bank, https://boris.unibe.ch/6734/1/LDR2010_Mapping.pdf (accessed 13 July 2025).

World Bank. 2017. Lao People's Democratic Republic: systematic country diagnostic – priorities for ending poverty and boosting shared prosperity. Washington D.C.: World Bank: https://openknowledge.worldbank.org/entities/publication/458c63ba-eb3d-55c6-aeb8-c1366479bf73 (accessed 8 January 2026).

World Bank. 2021. Consultation: country partnership framework. In Round 1 Virtual Consultation Meeting on "The Lao PDR's priorities for ending extreme poverty and promoting shared prosperity" — a Systematic Country Diagnostic, Vientiane.

World Bank. 2023. International Development Association project appraisal document on a proposed credit in the amount of SDR 19 million (US$25.50 million equivalent) to the Lao Peoples' Democratic Republic for a power distribution project. Washington D.C.: World Bank, https://documents1.worldbank.org/curated/en/099051223145529934/pdf/BOSIB17be6586202618d0c1fb0866be8d71.pdf (accessed 13 July 2025).

Yeophantong, P. 2016. China's hydropower expansion and influence over environmental governance in Mainland Southeast Asia. In Goh, E. (ed.), *Rising China's Influence in Developing Asia*, 289–318. New York (NY): Oxford University Press.

CHAPTER 6

A Karen approach to grassroots state formation: Shaping territories of life

Saw John Bright[1], Diana Suhardiman[2], Casper Palmano[3] and Kim Geheb[4]

Abstract

This chapter focuses on grassroots state formation processes in Karen State prior to and after Myanmar's 2021 military coup. It illustrates how local communities together with civil society organisations created and claimed political spaces in natural resource governance through the establishment of a Salween Peace Park. Bringing to light the close interlinkages between place-based knowledge, customary rights systems, and localised power dynamics, it shows how the SPP manifests as a grassroots response to state territorialisation attempts. Building on theories of territorialisation, we focus on the shaping of 'territories of life' and how local communities' aspirations and Karen perceptions of self-determination serve as key political foundations for the park's formation.

Keywords: place-based knowledge, customary rights systems, natural resource governance, hydropower development, land grabbing, Myanmar

Introduction

Following the military coup in February 2021, Myanmar transitioned from a country with a certain degree of political instability into a country undergoing fully-fledged civil war. This period marked the beginning of a systemic dismantling of earlier state transition processes, including the removal of the National League for Democracy's (NLD) power by jailing many of its leaders, intense political conflict, and widespread armed struggle. Nationally, Myanmar society responded strongly, through widespread protests and demonstrations, and systematically through

[1] Karen Environmental and Social Action Network, Thailand and Wageningen University, The Netherlands.

[2] Royal Netherlands Institute of Southeast Asian Studies, The Netherlands.

[3] Karen Environmental and Social Action Network, Thailand.

[4] Landscape Alliance, Nairobi, Kenya.

the formation of the National Unity Government that aims to serve as a national governing unit opposed to the military junta. Due to widespread military offensives and violent attacks on civilians, however, Myanmar's democratic struggle has been forced to retreat to scattered pockets of resistance within and outside the country. Within the context of natural resource governance, the military coup has erased all possible entry points to link the discussion on resource federalism with the national peace process negotiation (BEWG, 2017; Buchanan et al., 2013; Suhardiman et al., 2017). The junta has used the current political turmoil as justification not only to delay a promised election in August 2023, but also to take over all functions of the government, including departments managing natural resources, such as mining, hydropower, and land management. This period of political turmoil has also put on hold some large-scale infrastructure projects (Fawthrop, 2019; Paoletto and Uitto, 1996) because the military government has not been able to guarantee the safety of project developers. Other projects incorporated under the Belt and Road Initiative, however, continued to move forward.

This chapter focuses on grassroots state formation processes in Karen State. We define grassroots state formation processes as counter-territorial strategies designed, on the one hand, to resist the territorialisation advances of a more powerful state; and on the other, the mobilisation of territory as an expression of identity and cultural life that contrasts with that of the dominant state. Combined, these assertions convey the aspirations of the resident community for political self-determination and challenge the primacy of the nation state. The chapter focuses on the Salween Peace Park (SPP), illustrating how local communities together counter-territorialise in response to the territorial ambitions of the central state. The chapter delivers two main arguments: (a) that the SPP is an exercise in juxtaposing local governance and self-determination approaches with the militarised and often violent territorialisation efforts of the Myanmar state; and (b) that the territories-of-life approach employed by the SPP, while sharing much in common with the discussion on hydrosocial cycles, broadens our understanding of how local-level governance systems can meld the distinction between the biophysical and the social without privileging water over other critical natural resources.

Methods

The first author conducted an in-depth case study research (Burawoy, 1991; Yin, 1994) from October 2020 and December 2021 in Karen State, Myanmar, focusing on how local communities along the Salween River view the formation of the SPP. We conducted a series of focus group discussions, followed by key informant interviews and in-depth semi-structured interviews with SPP committee members, village heads, village elders, and villagers. We link information and insights gathered from this field research with policy review and institutional analysis pertaining to land and water governance in Myanmar, and in relation to the SPP charter.

Territorialisation, counter-territorialisation and the territories of life

Territorialisation can be understood as the way in which some actors with the power to delineate territories are able to regulate access to, and exploitation of, natural resources (Sack, 1986; Vandergeest and Peluso, 1995). Scholars have also shown, however, that when it comes to state territorialisation, there will invariably be spaces where the reach of the state is weak or non-existent (Scott, 2009). These 'zones of anomaly' are "troublesome to the rationalizing, centralizing and individuating tendencies" (Sivaramakrishnan, 1999: 33) of the state, which then targets them, only to encounter resistance. These areas are frequently considered 'frontier spaces' ripe for development and exploitation, and rendering to capitalist relations of production (Barney, 2009). Their inhabitants are frequently (if not typically) indigenous minorities who need to be 'captured' and 'enclosed' into the territorial and ideological spectrum of the state to be administered, and economically and culturally standardised (Scott, 1998).

Territorialisation is necessarily scaled. At its most superficial, local areas are located within the much wider scope of the state's territorial ambitions. Scale serves to reinforce the power of the state and enables this power to be exercised at local levels. The ability to create and claim territories is not the monopoly of the state or dominant powerful actors, but rather is constantly emerging from ongoing power struggles between powerful and less powerful actors. Similarly, decision-making authority to govern natural resources is not taken as given, but rather understood as an ongoing process of contestation that is (re)shaped by actors' interests, strategies, and 'bundles of powers' to access and benefit from resources (Ribot and Peluso, 2009; Suhardiman et al., 2019), beyond formal territorial boundaries, responsibilities, and mandates. These processes of scaled contestation and territorial (re)making are highly fluid and dynamic (Swyngedouw, 1997).

The Burmese military (the Tatmadaw) "... is deeply imbued with the idea that since 1948 it alone has been responsible for holding the Union together, defeating its enemies – both internal and external – and saving the country from chaos" (Selth, 2018: 28). This is articulated in Myanmar's 2008 Constitution which remains—at least ostensibly—in force. The Tatmadaw's interpretation of its constitutional mandate is highly conservative—to the extent that ethnic minority movements—whether armed or otherwise—are seen to challenge national unity (Oo, 2020). In contrast, its interpretation of how it should fund and enrich itself is very liberal, and for a large number of its internal military actions would appear to be motivated by the quest for access to natural resources as well as paving the way for large agribusiness corporations, mining, and other large-scale infrastructural development within which it retains an interest (Boutry et al., 2017; Kenney-Lazar et al., 2023; Woods, 2018). Karen State is no exception. Here, state territorialisation attempts have manifested in widespread land- and water-grabs, regardless of how this might affect peace, stability, or security.

The ways in which localised communities respond to state territorialisation efforts are diverse. In Myanmar, given the violence that accompanies state territorialisation, various ethnic communities have taken up arms, and organised themselves into Ethnic Armed Organizations (EAOs), one of which is the Karen National Liberation Army (KNLA)—the military arm of the Karen National Union. During the State Peace and Development Council era (1997-2011), the establishment of Tatmadaw 'peace villages' (i.e., state compliant) in Karen State (and elsewhere) represented the forcible insertion of state space into Karen communal spaces. These nuclei of violent control had at their centre Tatmadaw military camps, which surrounded themselves with a buffer of (often forcibly relocated) villagers. In return for not burning their homes down or forcibly relocating them, villagers would agree not to assist insurgents, and to provide free labour to the camp on a rotating basis (Scott, 2009: 180).

'State repelling' responses in this case normally comprised flight "...to inaccessible areas, scattering and breaking up into smaller and smaller groups, and pursuing subsistence techniques that [were] invisible or unobtrusive" (Scott, 2009: 180-181)—besides escaping to refugee camps across the border in Thailand. In more stable environments, people may "resist and subvert hegemonic control and attempt to keep governing forces at bay through often subtle acts, including evasion and foot dragging" (Hathaway, 2013: 79). Whatever the case, these various responses together comprise counter-territorialisation strategies (Lestrelin, 2011).

The idea of 'territories of life' may be defined as the territories and areas conserved by indigenous peoples and local communities. They are as diverse as the peoples and communities who shape and sustain them through their unique cultures, governance systems, and practices (ICCA Consortium, 2021). Territories of life share in common the following characteristics. Firstly, a close and deep connection between a territory or area and an indigenous people or local community. This relationship is generally embedded in history, social and cultural identity, spirituality, and/or people's reliance on the territory for their material and non-material wellbeing. Secondly, the custodian people or community makes and enforces decisions and rules (e.g., access and use) about the territory, area, or species' habitat through a functioning governance institution. Thirdly, the governance decisions and management efforts of the concerned people or community contribute to the conservation of nature (ecosystems, habitats, species, natural resources), as well as to community wellbeing (ICCA Consortium, 2024).

The idea of territories of life shares much in common with that of the hydrosocial cycle, the "…socio-natural process by which water and society make and remake each other over space and time" (Linton and Budds, 2014: 170)—but for the fact that the territories of life do not privilege water over and above the other processes in a landscape needed to sustain life. Water is but one (albeit essential) ingredient in the mix of (social/natural) things and (social/natural) processes that constitute the living territory. In the territories of life, the human inhabitants convene and moderate the relation between the biosphere and human life, and the humanly devised managerial, knowledge, and governance inputs needed to sustain biophysical life. The relation between these two domains is sufficiently blurred as to make them indistinguishable in a tightly integrated and circulating social-ecological system in which biological conservation is fused with sustainable living.

Enabling the land grab

Following the transition to a civilian, semi-democratic government in 2011, the Myanmar state initiated a series of legal land governance reforms, which controversially resulted in laws that enabled land dispossession (Kenney-Lazar et al., 2023; Suhardiman et al., 2019). Prior to the coup, state territorialisation attempts to control access to land were most apparent from controversies in Myanmar's legal land reforms. In 2012, two laws—the Vacant, Fallow, and Virgin (VFV) Land Management Law and the Farmland Law—were passed, which facilitated private land investment and ownership. Vacant, fallow, and virgin lands were land that had not been classified as agricultural, residential, or forest lands, regardless of how they were used.

Both laws threatened the land tenure security of smallholder farmers and customary land users while favouring larger land investors (Boutry et al., 2017; FIDH 2017; Linn 2015). VFV land users often did not have the resources to apply for a permit, and thus their land was vulnerable to acquisition by investors who did. Additionally, land registration to acquire a long-term land use certificate ('Form 7') did not necessarily provide a secure land tenure, as there were instances in which land had been confiscated from those who did possess this document (LIOH 2015). In Karen State, both the VFV Land Management Law and the Farmland Law would have classified all local community land, including ancestral lands, as unregistered VFV and farmland, and therefore belonging to the state. Lacking any formal legal framework to protect farmers' land, land grabbing was not only institutionalised in the government's bureaucratic rules and mechanisms, but legitimised.

Land governance reforms in Myanmar's ethnic states cannot be discussed and analysed in isolation from past debates on peace processes, decades of armed conflict, the movement of refugees, and the violence that forced people to abandon their farmland. In turn, these have set the pretext for the previous military government to grab significant amounts of farmland and communal forests, while furthering the scope and intensity of resource conflicts throughout Myanmar's various ethnic states. This highlights both the economic and political importance of exercising control over land in conflict areas (Kramer 2015; Linn 2015; Woods 2011, 2019). In line with the discussion on peace processes, discussions on political transition were taking place within the framework of the Nationwide Ceasefire Agreement (Dukalskis 2017; Huang 2013) signed by the former (military) President Thein Sein and eight representatives of various EAOs, including the KNU, in 2015 (Kramer, 2012, 2015; Kurlantzick, 2015).

Since the 2021 coup, the military junta has escalated its land grabbing, using it not only to exert its control through large-scale land concessions, but also to forcibly claim frontier spaces as new territory, through violent armed conflict.

Water grabs through hydropower development

Of the five dams planned for construction on the Salween mainstream in Myanmar, one (the Hatgyi) is in Karen State (see Fig. 6.1). If it proceeds, the Hatgyi will be constructed to the southeast of the SPP. Together or in isolation, the Salween dams will alter the river ecosystem and monsoon water level fluctuations, impacting a wide range of agricultural landscapes upon which local communities depend for their livelihoods (Magee and Kelley, 2009).

Dams and other hydraulic infrastructure are exceptional instruments of territorialisation, creating engineering bridgeheads into territories that the state deems important for the country's development or simply to assert its political power. They are "gigantic manifestations of state power, [that] can be seen as visualizing no less than a nation's dream of centralizing and controlling its resources and demonstrating it to the rest of the world" (Werner, 2014: 126). In much of Southeast Asia (including eastern Myanmar) hydropower has been used to project state power into geographies dominated by ethnic minorities (Bakker, 1999; Barabantseva, 2009; Middleton, 2022). According to Rogers et al. (2023: 6), "Hydropower infrastructure is largely understood as enabling the extension of state control over sometimes unruly, remote or marginal landscapes and their residents".

Dam development is usually accompanied by the construction of access roads into natural resource territories previously unavailable to the state, accompanied by new power lines that serve much the same purpose. The same can be said

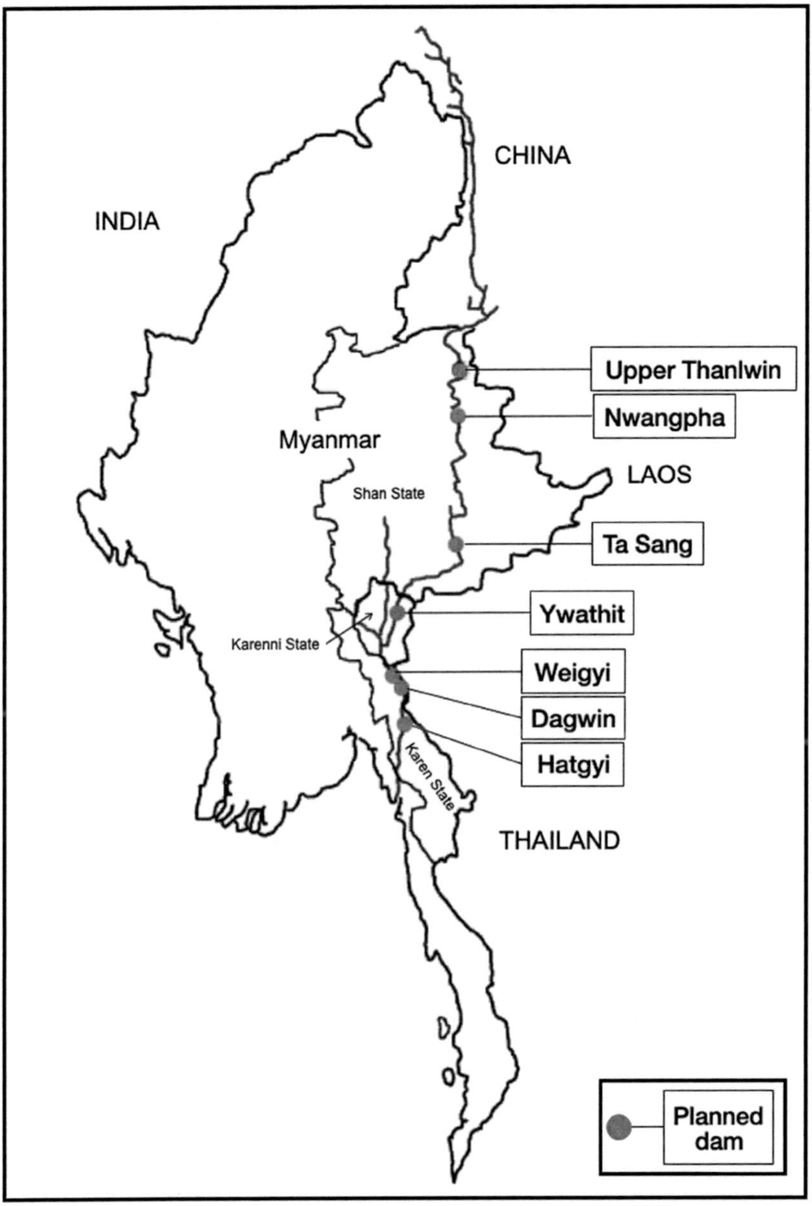

Figure 6.1: Map of Myanmar, showing the course of the Salween and planned mainstream hydropower dam sites in Shan, Karenni, and Karen states (Source: authors).

regarding camps for workers guarded by the military, or resettlement schemes to move 'project-affected people' out of the reservoir inundation areas. By compressing people into uniform settlements, the state can regulate these spaces and project its power further. Dams serve and are legitimised by modernity narratives, announcing the progression of the nation through the manifestation of the developmentalist state. The climate crisis reinforces this narrative, as the electricity produced from hydropower dams is often labelled as 'clean, green' energy.

There has been much violence associated with the Hatgyi Dam site, which has, in turn, forced large numbers of people to seek refuge from the violence elsewhere (Suhardiman et al., 2017). The violence that is frequently associated with hydropower development is well documented; hydropower, however, is but one tool of territorialisation—of which there are many others, such as roads, pipelines, railways, etc. From the state's perspective, these infrastructural interventions have a singular and powerful advantage: they can be legitimised as tools in the quest for national development.

Hydropower does lay claim to water, but there are also other ways in which this can occur. Prior to the 2021 coup, water management in Myanmar was supposed to be coordinated via an inter-ministerial body, the National Water Resources Council (NWRC). The NWRC is the product of the Union Government, having first been constituted in 2013 under the ultra-nationalist, pro-military Union Solidarity and Development Party government. Its work has included the formulation of the National Water Policy and the National Water Framework Directive. For all of these and any future water law, Integrated Water Resources Management (IWRM) is the stated central paradigm (Götz, 2019), while lamenting (as it does) that there is "little or no knowledge about [IWRM in Myanmar] except in the domain of water professionals. The whole country needs to be aware of IWRM principles and participatory approach" (NWRC, 2015: 2). 'Integration' would appear—at least implicitly—to be seen as centralising, with the NWRC calling for itself to have a major role in all water matters—assuming not only that the Union Government has the right to make this claim, but that ethnic states will comply.

The *kaw* system as grass-roots state building

Karen communities' access to their ancestral lands is structured and guided by the kaw system. According to both Farmland and VFV Laws, however, kaw lands do not 'belong' to anyone because they are not formally registered, making them vulnerable to land grabbing by the state or private interests (Cole, 2020). In any case, even if land titles were held, these would often be insufficient to stop a large-scale land grab. Karen communities in Mutraw District therefore rely on the kaw system to challenge these state interests, and the laws that support their territorial ambitions. The kaw system is specifically identified in the SPP's Charter (see below), and is also central to KNU land policy, where it is identified and considered under Article 3.3 (KNU, 2015).

A kaw territory comprises the land, water, and other natural resources within specific ancestral domains. Perceived and conceptualised as 'territories of life', a kaw territory may contain several different types of forest and land use. All kaw units are connected to each other, which, combined, belong to a specific kaw community. Each kaw is, therefore, unique, informed by its own specific context and history, but it never operates in isolation from other units. It is not uncommon to find different groups of local communities belonging to multiple kaw working together and adopting rules, belief systems, conservation, and natural resource practices. Kaw are also spiritual territory—in effect, all Karen land is 'owned' by the spirits, from whom the first person wishing to exploit the land must ask permission through a series of rites. The importance of spirits in the kaw system has led Cole (2020) to refer to it as a form of 'spectral sovereignty'. In effect, the spirits then cede user rights to the land, which the pioneer can then hold in perpetuity, and can pass on to his children. Once the spirits have granted permission for the land to be used, this pioneering use-rights holder can lend the land out to others, in a complex system of distributing use rights to those who need access to land (Cole, 2020). This system of user rights is distinctly Karen, and the kaw system represents a repository of indigenous Karen beliefs and practices that reflect their cultural identity and sets them (markedly) apart from the kinds of land user rights approaches espoused by the VFV Land Management Law and the Farmland Law.

Unlike individual land titles and ownership, the kaw governance system does not stand in isolation from wider natural resources. Here, humans maintain social relations and mutual responsibilities towards the nonhuman, and it is unthinkable to consider land as a material object of control and exclusive jurisdiction (Paul, 2018). The kaw system can be understood in multiple ways: as a management and governing system, a social institution, a social framework, and/or as spiritual and physical territory, depending on the context in which it is placed and connected with. The kaw system is highly dynamic, reflecting the complexity of the SPP landscape, and integrating indigenous ecological knowledge, protected areas, rotational swidden agriculture fields, and peaceful conflict resolution mechanisms.

The Karen tie this understanding of the natural resources on which they rely with cultural survival. As one proverb states, "mother told us to save yam species, father told us to save taro species. When we preserve up to 30 kinds, we will not perish when famine comes" (KESAN, 2015). Another proverb specifies that: "If you eat something from water, do not eat it together with something from the land" (KESAN, 2015), thus centring both cultural survival and biodiversity in kaw land

and water management. This way, Karen communities balance resource use from both land and water, instead of using it intensively and at the same time. At special events (e.g., during the harvest) local communities will eat fish, other aquatic animals, and animals reared on land, celebrating the moment that food has been secured.

The presentation and positioning of the SPP as 'territories of life', rooted in environmental integrity and cultural survival, is representative of Karen aspirations for peace and self-determination. The SPP embodies Karen communities' close relationship with the land as a system of reciprocal relations and obligations. As a local inhabitant stressed: "land is the foundation: as a nation, we need our land, both for livelihoods and cultural identity" (Interview, December 2020). Here, land is viewed not only as an economic asset or a source of livelihood alone, but is rather understood as a social landscape of profound socio-cultural, political, and spiritual importance and connection to the people who live on it. A local community member explained how, with the disruptions from military occupation and displacement, it is not possible to properly pass on these traditions and rituals:

> Our greatest desire is that the Burmese soldiers would leave so that we can once again practise the customs and traditions of our ancestors properly and in peace, in freedom and with our right of self-determination. When we are displaced, we can no longer practise our livelihoods, traditions, and customs properly (Interview, November 2020).

Here, the SPP embodies the idea of self-determination, as a territory of life where indigenous Karen communities govern their own affairs in their own territories, free from coercion and control by others.

The Salween Peace Park

The SPP was established in 2018 in Mutraw District of Karen State[5] and "...articulates a local vision for peace" (Paul et al., 2023: 274). The idea of 'peace parks' has particular relevance in Myanmar, juxtaposing with the violence wrought by the Tatmadaw and the military junta.

A key actor in the park's formation has been the Karen Environmental and Social Action Network (KESAN), a community-based, non-governmental organisation that works closely with the Karen National Union (KNU). The formation of the SPP is a complex endeavour rooted in KESAN's and local communities' strategies

[5] In this chapter, we use Karen place names. Thus, Karen State is what the Union Myanmar Government refers to as 'Kayin State', and Mutraw as 'Papun District'.

to develop "...alternative modes of possessing the earth, building a federal union and evoking 'the spirit of self-determination'" (Cole, 2020: 210). Here, conservation and the assertion of different ways of living are cast as part of the nationwide struggle—an act of resistance that works in tandem with other KNU strategies, that combined seek to counteract the state's territorial ambitions.

The establishment of the SPP followed the meeting of 300 local Karen leaders in May 2016. As an attempt to move the then discussion on peace processes forward with the NLD government, these leaders proposed the establishment of the park as an alternative way to navigate through the proposed dam developments on the Salween River and ongoing armed conflict (Suhardiman et al., 2017). Located in the area where armed conflict continues, the park is meant to provide space for sustainable rural livelihoods, environmental protection, wildlife conservation, and the needed resources for post-conflict recovery.

The SPP is driven by three main principles: (a) peace and self-determination; (b) environmental integrity; and (c) cultural survival. These principles reflect and are rooted in the kaw system as follows by acknowledging water, land, and the ecosystem as an integral part of people's livelihoods. This is most apparent in the context of forest governance. As one farmer whom we interviewed said: "If we have forest, we have water. And if we have water, we have life" (Interview, month and year). The SPP Charter (Article 118) phrases this as: "our worldview highly values the interdependence of the forest, land, water, wildlife, and human life. The people of SPP shall respect and value animist traditions, rituals, and ceremonies that are part of the collective memory of the Karen indigenous people."

The principles are guided by the SPP Charter, developed through community consultations, and ratified by popular referendum in 2018. Implementation of the charter is guided by Karen CSOs in general and the KESAN in particular. In 2018, 75 per cent of the inhabitants of Mutraw District voted to support the SPP charter that was drawn up in continuous consultation with communities from the SPP's 26 village tracts, representing approximately 60,000 inhabitants. In 2021, communities from outside the SPP's original boundaries petitioned to join the SPP, with communities from another nine village tracts comprising circa 13,000 inhabitants, voting to support the charter, and formally joining the SPP in December 2022.

The SPP has also expanded its area to include those lands around the proposed Hatgyi Dam site on the Salween. These include community ancestral land. By incorporating this land into the SPP's territories of life, it seeks to further counter-territorialise the state's hydropower-related territorial ambitions. Similarly, the SPP represents the institutional embodiment of KNU Land Policy and its objective to protect farmers' farmlands from land grabbing.

Institutionally, the park is managed by a General Assembly as its governing body, within which community representatives hold the majority and are elected every

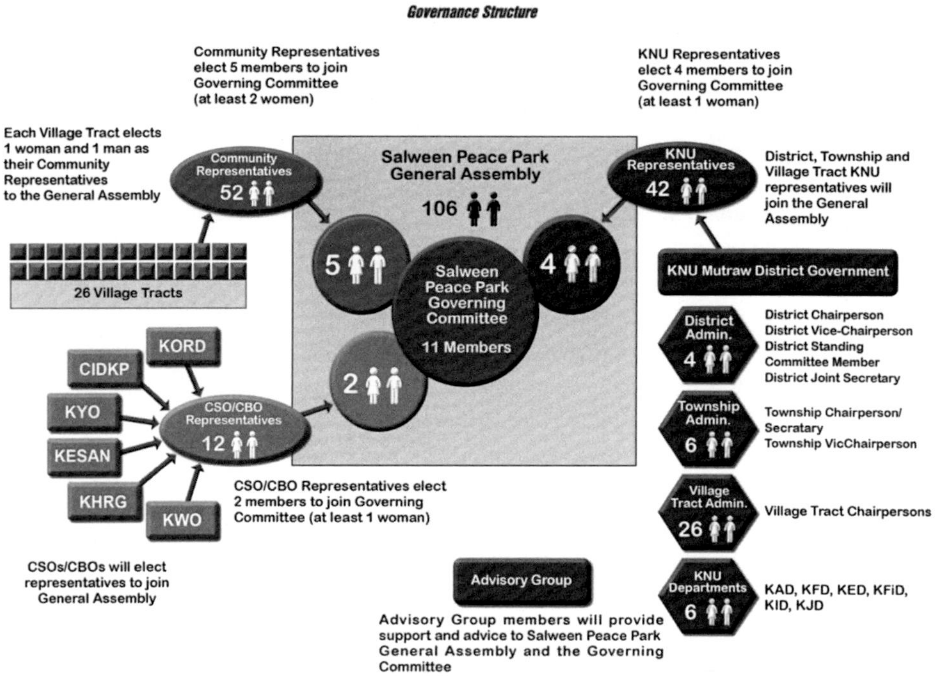

Figure 6.2: Salween Peace Park governance structure (adapted from KESAN, 2017). Notes: each participating group elects members to the SGC. For the CSO/NGO and District Government groups, at least one woman must be amongst the elected representatives; for the Community Representatives group, at least two women must be amongst the elected representatives. CBO = Community-based Organisation. CSO = Civil Society Organisation; GA = General Assembly; SGC = SPP Governing Committee (Source: KESAN, 2017).

three years through popular vote (Fig. 6.2). Through the General Assembly, the SPP incorporates the more than 100 kaw systems within its borders to holistically govern the area and ensure sustainable livelihoods and environmental conservation. Besides local community representation, CSOs and KNU representatives are also part of the Assembly. This hybrid governance structure and bottom-up approach allows larger landscape-scale decisions taken by the General Assembly to remain flexible and adaptable while ensuring that grassroots aims and techniques are not lost in macro-level planning. It also ensures local communities' ownership of the General Assembly's decision making, leading to effective implementation at ground level by the SPP's various kaw communities. Furthermore, this approach allows the SPP to further build upon local community sustainable water and land governance.

At present, the SPP incorporates 73,416 acres of community forest, including a heavily forested area along the border between Myanmar and Thailand, through which the Salween River flows. Different areas of community forest have been classified into community conserved, preserved, and managed forest. While this classification is important to ensure that local residents are well informed about and committed to forest preservation in line with KNU Forest Department's policy and guidelines, community forest can only exist and function when it is incorporated within the kaw governing structure and thus forms an integral part of the latter.

The SPP approaches the debate on self-determination from the perspective of what this means for indigenous Karen communities on the ground. As an SPP village elder said: "The perspectives on self-determination by political elites who for decades have debated this issue at the very top overlook local communities' definitions, expectations, and reasons for self-determination" (Interview, December 2020). The SPP does not lay claim to the territory of indigenous Karen communities by means of secession; rather, it seeks the day-to-day management of their ancestral land as territories of life. Together with the KNU and Karen CSOs, the SPP aims to empower local communities and institutions and strengthen the interconnection between people, culture, and nature in their (own) territories.

Following the February 2021 military coup, the Tatmadaw launched an airstrike against villagers in Mutraw District and surrounding areas, killing at least 19 people, injuring at least 27 more, and forcing several thousands to flee their villages (Cowan, 2023). Violent armed conflict between the Tatmadaw and EAOs in Karen and Karenni states has significantly hampered community abilities to safeguard the SPP's ecosystems. Moreover, monitoring resource extraction in many parts of the park has been complicated by the security threat, and the growing number of displaced people that are now dependent on forest resources to survive. Since the coup, some 100,000 people have sought refuge within the SPP.

Nonetheless, these challenges do not stop local community and Civil Society Organizations' (CSOs) efforts and determination to continue with the SPP. They have responded to these challenges in several ways: first, they prioritise support for affected local communities while trying to sustain ecosystem values as much as they can. This means allocating some forest conservation areas to local communities who have lost their own land and farmlands. Second, they strengthen and broaden their networks to also include medical organisations (e.g., Doctors Without Borders) as well as refugee organisations so as to better support affected local communities. This way, the SPP's networks have expanded to include medical centres in surrounding urban areas as well. Last but not least, together with seven Karen indigenous groups in Taw Oo, KESAN has started a new initiative to form a similar park: Thawthi Taw Oo indigenous park. The park is currently in the initial stages of formation and is inspired by the SPP.

Placing the SPP within the broader context of Myanmar's current political struggles, the park presents an alternative proposition to that expressed by successive Myanmar military regimes and, indeed, the NLD-led government. The SPP reflects not only local communities' views on natural resource governance, but also their perceptions and contributions to shaping Myanmar's wider political landscape. Karen communities understand that "[s]pace does not exist in itself, it is produced" (Schmid, 2008: 28), and therefore seek to redefine their territory away from the space of control demanded by the state, to one of engagement. Local communities view territory as a system of thought, allowing them to shape it for a variety of engagement, creative, relational, and integrative purposes, to channel and convey their needs and aspirations, to enact their idea of citizenship, and push for more accountable state–citizen relations.

Taking advantage of scale

At its lower scales, the concepts and approaches embodied in the SPP require acceptance by those living in Mutraw District which, as we have seen above, has been obtained. The park has wide support internally because it does contrast with the land-grabbing approaches espoused by the Union Government and its military. During the last meeting convened by the KNU's Executive Committee, the Salween Peace Park was accepted into the KNU legal fold as official policy. The kaw system is recognised by the KNU and embodied in its legislation and policy. As Hong (2017) has argued, by writing kaw into law, these assemblages of kaw land communities become, in effect, 'scalar units' thereby becoming a legible and scalable element of local governance. In turn, the expression of Karen land law also has scale implications and dimensions. By expressing the law, the KNU pronounces state legitimacy by deploying the same 'tools of state' that any country would, irrespective of whether or not its laws are recognised by the Union Government. In this way, its state-forming strategies are fundamentally osmotic, for they push outwards against, and into, the pressures applied by the central state. Here, the configuration of the Karen State emerges not just through the defence of the boundaries of state, but by mobilising culture, everyday practices, identity, and recognition (see below) in an integrated process that seeks to establish presence.

The SPP has 'leapfrogged' the Union Government—it is not a scalar avenue that it seeks to pursue in its legitimacy and recognition efforts, regarding this more as a dead end (for now) in its pursuit of sovereignty. Instead, it has sought to deploy wider instruments of scale from international initiatives. Thus, 'living landscapes' emerge from the ICCA[6] Consortium, of which KESAN is an active member; label-

[6] "ICCA sounds like an acronym, but it is not. It is an abbreviation for 'territories and areas conserved by indigenous peoples and local communities'" (ICCA Consortium, 2024).

ling the SPP as a 'peace park' and as an Indigenous Community and Conservation Area derives from IUCN-facilitated ideas; actors working to further the SPP engage in protests against dam development and thereby connect to a network of localised—but also global—anti-dam movements (Götz, 2019).

Similarly, opposition to the Hatgyi and other Salween dams has obtained support from International Rivers, and "a network of community leaders and [CSOs] have organized under the name 'Save the Salween Network' who have regularly issued statements stating that the Salween River is the lifeblood of local people, providing livelihood opportunities and links with cultural identity" (Bright, 2019: 80). By reaching into international arenas, KESAN and the KNU seek to gain recognition for the SPP from these scales, in an effort to countermand the alternative identities that the Union Government seeks to impose.

Cole (2020: 198) has characterised KESAN's approach to scales as "the pragmatic politics of messy translations and constant messing with scales." In this sense, KESAN's activities are highly adaptive—identifying new (scaled) opportunities and seeing what dividends these might yield as well as seeking recognition in a multiplicity of arenas. This 'messing with scales' is, once again, motivated by juxtaposition—of contrasting the peaceful and sovereign ambition of the Karen state against the violence wrought by the Tatmadaw and its territorial claims. So too, this approach confounds the implied morality of the Tatmadaw's position that it is simply trying to keep the Union together.

Conclusion

The formation of the SPP illustrates a grassroots counter-territorial effort to the state's own formation processes. By formulating its own land policies—built around systems of land and water management that have been established for generations—the Karen assert their own system for living and being. In practice, land, water, living, and culture intermingle in ways that are challenging to separate, and while this intermingling has much in common with the hydrosocial cycle, it does not privilege one human or natural resource over others. The intense integration inherent to the kaw system is in itself a defence against the fragmenting processes applied by the state in their water or land or energy or forest policies that are, in turn, designed to make ethnic lands available and amenable to its own interests—in particular, with which to fund its military, enrich its military firms, and reward its cronies (Woods, 2018).

Local communities and CSOs socio-politically construct and materialise their belonging through the kaw system, and through the SPP. The SPP's establishment and resilience are rooted in Karen communities' cultural identity, in turn embedded in the kaw governance structure as indigenous territories of life. Here, local communities, CSOs, and other state actors make and claim territories not to exert political power and interests, but as a defence against the state-derived threats to their livelihood sustainability, the preservation of their cultural identity, and their pursuit of political self-determination. By asserting these rights and 'ways of living', the Karen create an osmotic pressure from within their state that presses outwards in defiance of the state.

These ways of living juxtapose with the violence associated with the state's own territorialisation efforts. Juxta positionality also emerges in KESAN's own 'messing with scales' as it seeks recognition at international levels that also legitimises its position and territorial rights.

Illustrating how state formation processes are embedded in grassroots actors' strategies embedded in integrated life–land–water-environmental relations offers an alternative view and perspective of what 'the state' could and should entail in the context of natural resource governance. It questions the notion of the nation state in the context of Myanmar's military dictatorship and reveals multiple grassroots forces and inter-scalar power dynamics that continuously (re)shape state–people relations. These forms are neither fixed nor bound within the formal boundary of the sovereign state; rather, they are based upon the dialectic relation of rights and authority (Lund, 2016). Lund notes that authority and rights are co-constitutive and conceptually tied together by recognition. Thus, "when institutions recognize claims to property or citizenship, they themselves become recognized by the claimants of these rights" (Eilenberg and Lund, 2017: 1). Mutual recognition, then, constitutes a contract that links citizenship rights to authority, and vice versa. This dialectic relation of rights and authority explains how less powerful actors create political spaces of engagement even within the realm of extreme authoritarianism and violence. Local communities, CSOs, and other state actors shape the process of state formation through their alliances and mutual recognition of how belonging and rights relate to their political authority and vice versa.

Local communities' counter-territorialisation strategies are not without limitation, especially in the context of violent armed conflict and struggles. Nonetheless, our case study also shows local actors' resilience to endure conflict, while relying on ever-growing solidarity and the power of the collective.

References

ADB (Asian Development Bank). 2012. Myanmar: energy sector initial assessment. October 2012. Manila: ADB, https://www.adb.org/sites/default/files/institutional-document/33719/myanmar-energy-sector-assessment_0.pdf (accessed 12 September 2023).

Bakker, K. 1999. The politics of hydropower: developing the Mekong. *Political Geography* 18(2): 209–232, https://doi.org/10.1016/S0962-6298(98)00085-7.

Barabantseva, V.E. 2009. Development as localization: ethnic minorities in China's official discourse on the Western Development Project. *Critical Asian Studies* 41(2): 225–254, https://doi.org/10.1080/14672710902809393.

Barney, K. 2009. Laos and the making of a 'relational' resource frontier. *Geographical Journal* 175(2): 146–159, https://doi.org/10.1111/j.1475-4959.2009.00323.x.

BEWG (Burma Environmental Working Group). 2017. Resource federalism: a roadmap for decentralized governance of Burma's natural heritage, https://www.burmalibrary.org/sites/burmalibrary.org/files/obl/docs23/BEWG-2017-ResourceFederalism-en-red.pdf (accessed 12 June 2024).

Boutry, M., Allaverdian, C., Mellac, M., Huard, S., Thein, U.S., Win, T.M. and Sone, K.P. 2017. *Land tenure in rural lowland Myanmar: from historical perspectives to contemporary realities in the Dry Zone and the Delta.* Of Lives and Land Myanmar Research Series. Yangon: GRET, https://gret.org/en/publication/land-tenure-in-rural-lowland-myanmar-from-historical-perspectives-to-contemporary-realities-in-the-dry-zone-and-the-delta/ (accessed 12 June 2024).

Bright, S.J. 2019. Rites, rights, and water Justice in Karen State: a case study of community-based water governance and the Hatgyi Dam. In Middleton, C. and Lamb, V. (eds.), *Knowing the Salween River: resource politics of a contested transboundary river*, 71–86. Cham: SpringerOpen.

Buchanan, J., Kramer, T. and Woods, K. 2013. Developing disparity: regional investment in Burma's borderlands. Amsterdam: Transnational Institute – Netherlands Burma Centre, https://www.tni.org/files/download/tni-2013-burmasborderlands-def-klein-def.pdf (accessed 25 March 2024).

Burawoy, M. 1991. The extended case method. In M. Burawoy (ed.), *Ethnography unbound: power and resistance in the modern metropolis*, 271–287. Los Angeles (CA): University of California.

Cole, T. 2020. Possessed earth: ownership and power in the Salween Peace Park of Southeast Myanmar. PhD Thesis. Department of Social Anthropology, Stockholm University, Stockholm, https://su.diva-portal.org/smash/get/diva2:1492910/FULLTEXT01.pdf (accessed 21 September 2023).

Cowan, C. 2023. Award winning, indigenous peace park dragged into fierce conflict in Myanmar. *Mongabay* 15 May 2023, https://news.mongabay.com/2023/05/award-winning-indigenous-peace-park-dragged-into-fierce-conflict-in-myanmar/ (accessed 3 January 2024).

Dukalskis, A. 2017. Myanmar's double transition: political liberalization and the peace process. *Asian Survey* 57(4): 716–737, https://doi.org/10.1525/as.2017.57.4.716.

Eilenberg, M. and Lund, C. 2017. State formation through the production of property and citizenship. Policy Brief. Copenhagen: Copenhagen Centre for Development Research.

Fawthrop, T. 2019. East Asia's last undammed river. *The Diplomat*, 13 June 2016, https://thediplomat.com/2016/06/east-asias-last-undammed-river/ (accessed 24 March 2024).

FIDH (Federation Internationale pour les Droits Humains). 2017. Land of sorrow: human rights violations at Myanmar's Myotha Industrial Park, https://www.fidh.org/IMG/pdf/fidh_report_-_land_of_sorrow_-_human_rights_violations_at_myanmar_s_myotha_industrial_park.pdf (accessed 29 March 2026).

Götz, J.M. 2019. Contested water governance in Myanmar/Burma: politics, the peace negotiations and the production of scale. In Middleton, C. and Lamb, V. (eds.), *Knowing the Salween River: resource politics of a contested transboundary river*, 87–106. Cham: SpringerOpen.

Hathaway, M.J. 2013. *Environmental winds: making the global in Southwest China*. Berkeley (CA): University of California Press.

Hong, E. 2017. Scaling struggles over land and law: autonomy, investment and interlegality in Myanmar's borderlands. *Geoforum* 82, 225–236, https://doi.org/10.1016/j.geoforum.2017.01.017.

Huang, R.L. 2013. Rethinking Myanmar's political regime: military rule in Myanmar and implications for current reforms. *Contemporary Politics* 19(3): 247–261, https://doi.org/10.1080/13569775.2013.804149.

ICCA Consortium. 2021. Territories of life: 2021 report. ICCA Consortium, https://report.territoriesoflife.org/ (accessed 10 March 2024).

ICCA Consortium. 2024. Territories and areas conserved by indigenous peoples and local communities, https://www.iccaconsortium.org/discover/ (accessed 27 June 2024).

Kenney-Lazar, M., Suhardiman, D. and Hunt, G. 2023. The spatial politics of land policy reform in Myanmar and Laos. *Journal of Peasant Studies* 50(4): 1529–1548, https://doi.org/10.1080/03066150.2022.2054700.

KESAN (Karen Environmental and Social Action Network). 2015. Water relational values of Karen indigenous communities in the Salween Peace Park. KESAN report to USIP.

KESAN (Karen Environmental and Social Action Network). 2017. The Salween Peace Park Program, https://kesan.asia/salween-peace-park-program-3936/ (accessed 13 September 2024).

Kiik, L. 2016. Nationalism and anti-ethno politics: why 'Chinese development' failed at Myanmar's Myitsone Dam. *Eurasian Geography and Economics* 57(3): 374–402, https://doi.org/10.1080/15387216.2016.1198265.

KNU (Karen National Union). 2015. Land Policy. Office of the Supreme Headquarters, Karen National Union, Kawthoolei, December 2015, https://knuhq.org/admin/resources/policies/pdf/knu_land_policy_eng.pdf (accessed 16 February 2024).

Kramer, T. 2012. Ending 50 years of military rule? Prospects for peace, democracy and development in Burma. Oslo: NOREF, https://ciaotest.cc.columbia.edu/wps/noref/0027052/f_0027052_25454.pdf (accessed 5 March 2024).

Kramer, T. 2015. Ethnic conflict and land rights in Myanmar. *Social Research* 82(2): 355–374, https://themimu.info/sites/themimu.info/files/documents/Academic_article_Ethnic_Conflict_and_Land_Rights_Tom_Kramer_2015.pdf (accessed 3 March 2023).

KRW (Karen Rivers Watch). 2016. Karen State September 2016 conflict: the real motivations behind renewed war, https://www.burmalink.org/karen-state-september-2016-conflict-the-real-motivations-behind-renewed-war/ (accessed 1 June 2024).

Kurlantzick, J. 2015. Myanmar's ceasefire deal comes up short. *The Diplomat* 11 October 2015, https://thediplomat.com/2015/10/myanmars-cease-fire-deal-comes-up-short/ (accessed 19 August 2023).

Lestrelin, G. 2011. Rethinking state-ethnic minority relations in Laos: internal resettlement, land reform and counter-territorialization. *Political Geography* 30(6): 311–319, https://doi.org/10.1016/j.polgeo.2011.06.005.

Lin Htet Myat, 2021. Myanmar junta sells resources to neighbors in exchange for 'legitimacy'. *The Irrawaddy* 15 June 2021, https://www.irrawaddy.com/opinion/guest-column/myanmar-junta-sells-resources-to-neighbors-in-exchange-for-legitimacy.html (accessed 11 November 2023).

Linn, S.N. 2015. Myanmar: conflicts over land in a time of transition. *Economic and Political Weekly* 9, 69–73, https://www.jstor.org/stable/24481518 (accessed 14 February 2024).

Linton, J. and Budds, J. 2014. The hydrosocial cycle: defining and mobilizing a relational-dialectical approach to water. *Geoforum* 57(2014): 170–180, https://doi.org/10.1016/j.geoforum.2013.10.008.

LIOH (Land in Our Hands). 2015. Destroying people's lives: the impact of land grabbing on communities in Myanmar. Yangon: Land In Our Hands, https://www.tni.org/files/article-downloads/lioh_research_report_eng_0.pdf (accessed 11 December 2023).

Lund, C. 2016. Rule and rupture: state formation through the production of property and citizenship. *Development and Change* 47(6): 1199–1228, https://doi.org/10.1111/dech.12274.

Magee, D. and Kelley, S. 2009. Damming the Salween River. In Molle, F., Foran, T. and Käkönen, M. (eds.), *Contested waterscapes in the Mekong Region: hydropower, livelihoods and governance*, 115–140. London: Earthscan.

Middleton, C. 2022. The political ecology of large hydropower dams in the Mekong Basin: a comprehensive review. *Water Alternatives* 15(2): 251–289, https://www.water-alternatives.org/index.php/alldoc/articles/vol15/v15issue2/668-a15-2-10/file.

NWRC (National Water Resources Council). 2015. Myanmar National Water Policy. Yangon: NWRC, https://faolex.fao.org/docs/pdf/mya201049.pdf (accessed 28 March 2024).

Oo, N.Y. 2020. A new Tatmadaw with old characteristics. In Montesano, M.J., Chong, T. and Pračhak Kộngkīrati (eds.), *Praetorians, profiteers or professionals? Studies on the militaries of Myanmar and Thailand*, 30–69. Singapore: ISEAS-Yusof Ishak Institute.

Paoletto, G. and Uitto, J.I. 1996. The Salween River: is international development possible? *Asia Pacific Viewpoint* 37(3): 271–284, https://www.researchgate.net/publication/216535514_The_Salween_River_Is_International_Development_Possible (accessed 3 June 2024).

Paul, A. 2018. With the Salween Peace Park, we can survive as a nation: Karen environmental relations and the politics of an indigenous conservation initiative. Master's thesis. Toronto (ON): York University, https://yorkspace.library.yorku.ca/bitstreams/950e7a67-b3cb-48d5-94a7-0c084870ebd4/download (accessed 12 June 2024).

Paul, A., Roth, R. and Sein Twa, S.P. 2023. Conservation for self-determination: Salween Peace Park as an Indigenous Karen conservation initiative. *AlterNative: An International Journal of Indigenous Peoples* 19(2): 271–282, https://doi.org/10.1177/11771801231169044.

Ribot, J.C. and Peluso, N.L. 2009. A theory of access. *Rural Sociology* 68(2): 153–181, https://doi.org/10.1111/j.1549-0831.2003.tb00133.x.

Rogers, S., Fung, Z., Lamb, V., Gamble, R., Wilmsen, B., Wu, F. and Han, X. 2023. Beyond state politics in Asia's transboundary rivers: revisiting two decades of critical hydropolitics. *Geography Compass* 17(4):e12685, https://doi.org/10.1111/gec3.12685.

Sack, R.D. 1986. *Human territoriality: its theory and history*. Cambridge: Cambridge University Press.

Salween Watch Coalition. 2016. Current status of dam projects on the Salween River, https://river-resourcehub.org/wp-content/uploads/files/attached-files/salween_factsheet_2016.pdf (accessed 29 March 2026).

Scott, J.C. 1998. *Seeing like a state: how certain schemes to improve the human condition have failed.* Yale Agrarian Studies. New Haven (CT): Yale University Press.

Scott, J.C. 2009. *The art of not being governed: an anarchist history of upland Southeast Asia.* New Haven (CT): Yale University Press.

Schmid, C. 2008. Henri Lefebvre's theory of the production of space: towards a three-dimensional dialectic. In Goonewardena, K., Kipfer, S., Milgrom, R. and Scmid, C. (eds.), *Space, difference, everyday life: reading Henri Lefebvre*, 27–45. New York (NY): Routledge.

Selth, A. 2018. The defence services. In Simpson, A., Farrelly, N. and Holliday, I. (eds.), *Routledge handbook of contemporary Myanmar*, 25–34. London: Routledge/Taylor & Francis Group.

Sivaramakrishnan, K. 1999. *Modern forests: statemaking and environmental change in colonial Eastern India.* Stanford (CA): Stanford University Press.

Suhardiman, D., Kenney-Lazar, M. and Meinzen-Dick, R. 2019. The contested terrain of land governance reforms in Myanmar. *Critical Asian Studies* 51(3): 368–385, https://doi.org/10.1080/14672715.2019.1630845.

Suhardiman, D., Rutherford, J. and Bright, J. 2017. Putting violent armed conflict in the center of the Salween hydropower debates. *Critical Asian Studies* 49(3): 349–364, https://:10.1080/14672715.2017.1328284.

Suhardiman, D., Kenney-Lazar, M. and Meinzen-Dick, R. 2019. The contested terrain of land governance reforms in Myanmar. *Critical Asian Studies* 51(3): 368–385, https://doi.org/10.1080/14672715.2019.1630845.

Suhardiman, D., Scurrah, N. and Ayemyaing, N. 2021. Scalar politics, power struggles and institutional emergence in Daw Lar Lake, Myanmar. *Journal of Rural Studies* 87 (2021): 32–44, https://doi.org/10.1016/j.jrurstud.2021.08.023.

Swyngedouw, E. 1997. Excluding the other: the production of scale and scaled politics. In Lee, R. and Wills, J. (eds.), *Geographies of economies*, 167–176. London: Arnold.

Vandergeest, P. and Peluso, N. 1995. Territorialization and state power in Thailand. *Theory and Society* 24(3): 385–426, https://doi.org/10.1007/BF00993352.

Werner, H. 2014. Rivers, dams and landscapes: engaging with the modern on contested ground. In Nüsser, M. (ed.). *Large dams in Asia: Contested environments between technological hydroscapes and social resistance*, 125–148. Dordrecht: Springer.

Woods, K. 2011. Ceasefire capitalism: military–private partnerships, resource concessions and military–state building in the Burma–China borderlands. *Journal of Peasant Studies* 38(4): 747–770, https://doi.org/10.1080/03066150.2011.607699.

Woods, K. 2018. The conflict resources economy and pathways to peace in Burma. Peace Works No. 144. Washington (DC): United States Institute for Peace, https://www.usip.org/sites/default/files/2018-11/pw_144_the_conflict_resource_economy_and_pathways_to_peace_in_burma.pdf (accessed 3 July 2024).

Woods, K. 2019. Green territoriality: Conservation as state territorialization in a resource frontier. *Human Ecology* 47(2): 217–232, https://doi.org/10.1007/s10745-019-0063-x.

Yin, R.K. 1994. *Case study research: Design and methods.* New Delhi: Sage Publications.

CHAPTER 7

Placing and connecting the Mekong River Commission's sand discourses

Jean-François Rousseau[1] *And Tony Dinh*[2]

Abstract

This chapter delves into the Mekong River Commission's perspectives on sand resources, in the context of a growing regional sediment deficit and the ensuing consequences. The broader context of sand mining, its impacts, and the challenges associated with governing this 'development mineral' are explored. The analysis focuses on the Commission's position concerning two major drivers of sediment alterations, hydropower expansion and sand mining. We analysed 117 MRC reports published over two decades. The findings indicate that sand mining and related impacts emerged only recently in the MRC's narratives. Accordingly, numerous information gaps are acknowledged, although strategies are also emerging to promote the 'sustainable' management of this most important resource. Still, questions remain on the Commission's capacity to implement this ambitious programme.

Keywords: Sand mining, Hydropower, Sediment trapping, Mekong River Commission, Discourse analysis

Introduction: Sand and the Mekong Region

Writing about the politics of hydropower development in the Mekong Region in the late 1990s, Karen Bakker (1999: 218) explained that, in the Mekong River Delta, "decreased siltation and increased flooding are often-cited concerns" with respect to the downstream impacts of mainstream and tributary hydropower dams then planned in the Lao PDR. Over two decades later, many of these hydropower projects have materialised both in the Lao PDR and beyond, and the sediment discharge of the Mekong River has indeed undergone significant reductions, including in the delta (e.g., Hecht et al., 2019; Jordan et al., 2019; Le et al., 2020; Schmitt et al., 2017).[3]

1 University of Ottawa, Ottawa (ON), Canada.

2 University of Ottawa, Ottawa (ON), Canada.

3 The delta spans both Cambodian and Vietnamese territory (cf. MRC, 2019a, 2020, 2021). We use 'Vietnam Mekong Delta' when referring to scholarship that emphasises the Vietnamese part of the delta alone.

Hydrological research has also increasingly acknowledged sediment budget fluctuations as outcomes from a range of drivers that include not only sediment trapping by reservoirs, but also deforestation, climate change, sea level rise, and sand mining. This scholarship emerges alongside NGO and grassroots advocacy campaigns that increasingly highlight the diverse drivers of sediment budget diminutions in the Mekong Region, as well as their impacts on both riparian ecosystems and livelihoods (e.g., Marschke and Rousseau, 2022a; Mother Nature Cambodia, n.d.; WWF, 2018). As this unfolded, scholars and campaigners alike have called for the establishment of transboundary sand governance mechanisms, including within the Mekong River Commission (MRC) (Hirsch, 2020; Mercado and Goichot, 2023).

In this chapter, we aim to investigate the importance that sand holds in MRC discourses and assess whether this has evolved in light of increasing regional impacts from a growing sediment deficit. Building from MRC documents published over two decades, we focus on how the Commission has involved itself in two important anthropogenic activities driving significant sand budget changes, namely hydropower expansion and sand mining. We also delve into how sand connects with broader MRC discourses on resource management.

To achieve our aim, we first situate sand mining in general, and highlight impacts from sand budget changes and some of the circumstances that make this 'development mineral' (Franks, 2020) difficult to govern. We also approach the Mekong Region's centred hydropolitics scholarship and find that it has so far allocated little attention to sand. We build from this scholarship to situate our methodology emphasising institutional discourse analysis, an approach common in this literature. Our context section provides more details on the changes in sediment resource availability and factors driving these fluctuations in the Mekong Region. In our methodology, we present the corpus of MRC reports which we reviewed, together with the coding and analysis strategies that we deployed. Our data document the MRC's discourses on the causes and impacts of sediment budget changes and how these narratives have evolved and became increasingly frequent in the reports over the last decades.

Overall, we find that the MRC has only recently begun to allocate attention to decreasing sand budgets and their impacts. While the Commission has long stressed the influence of reservoir construction on sediment budgets, it has only recently started to approach sand and the factors shaping its decreased availability in a more exhaustive manner. As drivers of sediment reduction are increasingly scrutinised, implementing the solutions required to mitigate shifts in sediment budgets—including re-evaluating a development model that undervalues sand—appears increasingly challenging.

Sand scholarship: Sand mining and impacts

Silt, sand, and gravel all consist in natural aggregates, a category that encompasses materials of different diameters found in riverine, lake, coastal, marine, terrestrial/quarry, and desert environments or manufactured mechanically (e.g., by crushing rocks) (Langer, 2016; Valentine, 2019). According to UN estimates, the world consumed over 45 billion tonnes of non-metallic minerals in 2020, amounting to almost half of all material extracted that same year (UNEP, 2024). Construction aggregates make up most of the global demand for these minerals. While desert sand is improper for most industrial applications, river aggregates are widely used as key ingredients for asphalt and concrete. This is especially so in Global South countries where regulation overseeing river sand extraction is generally less stringent than in the Global North (Bravard et al., 2013; Koehnken et al., 2020).

Sand demand growth is particularly important in countries experiencing rapid urbanisation, as expanding infrastructure requires securing new sand procurement networks. Given sand's prevalence in the construction sector, global demand now exceeds natural regeneration rates (Bendixen et al., 2019), with consumption/natural supply imbalances being documented along Cambodian stretches of the Mekong River for instance (Hackney et al., 2021).

Sand mining-driven changes in riverine sediment budgets are complex and notably manifest in two important ways. Sand mining is first associated with rises in suspended sediment flows, as dredging and evacuating water overflow from sand barges lead to increased turbidity. This in turn impacts flora and fauna and leads to the reshuffling of heavy metals and other pollutants lying under the riverbed back to the surface (see Marschke and Rousseau, 2022b).

Second, the removal of sand diminishes the sediment flow available in a river system or along sections thereof. Impacts are first and foremost associated with increased water bank and infrastructure erosion and river channel incision (deepening). Sediment-free water, also known as 'hungry water', is more abrasive than its sediment-rich equivalent (which fosters sediment accretion). Hydropower expansion accentuates this process with aggregates being blocked from flowing downstream from reservoirs. For dam operators, reservoir sedimentation is positive, for it can prevent turbine damage, but it also diminishes reservoir operation efficiency (Li et al., 2022). Channel incision is understood as a desirable outcome for fluvial transport and flood control, but it also facilitates saline water intrusion in coastal areas such as the Mekong Delta (Bravard et al., 2013; Park et al., 2022).

Sand political economy

Natural aggregates consist in 'low value minerals' commanding a low price per weight unit (Franks, 2020). Sand is therefore most often extracted in the vicinity of construction sites, given that its bulkiness commands high transport prices (Da and Le Billon, 2022). Sand trade accounts for only a very minor portion of the global sand demand and long-distance exports such as those headed to Singapore remain a rarity (see Lamb et al., 2019). Likewise, sand extraction consists of a highly fragmented sector dominated by small-scale or artisanal operations, with multinational conglomerates controlling only modest market shares of the sector (Da and Le Billon, 2022).

As a 'development mineral', sand is predominantly extracted, processed, and utilised locally, in short-haul commodity circuits serving local economic needs, including those driven by the infrastructure development sector. Development materials, therefore, have a great potential to contribute to local development and poverty reduction initiatives (Franks, 2020). Sand's importance thereby transcends its sole monetary value, fostering both local job opportunities and infrastructure development—including dams—and typically yield greater livelihood benefits than export-oriented ores.

Conversely, these same circumstances potentially create sand governance and sustainability challenges. Emerging sand scholarship highlights the many problems that accrue in the sand mining sector. These include illegal (sometimes mafia-like) extraction activities, poor labour conditions, unsustainable mining practices, violence, corruption, extraction data deficits, inappropriate regulatory regimes at the local, national, and global levels, deficient and poorly implemented licensing processes, etc. (Beiser, 2019; Bikkina, 2017; Bisht, 2021; Bisht and Gerber, 2017; Lamb et al., 2019). The imperative for governments to secure the cheap sand procurement networks necessary to achieve development objectives, together with clientelist relations, could also explain why authorities often accord low scrutiny to the sand sector and why sand is often considered a 'sensitive' topic, including in the Mekong Region (e.g., Sopera, 2022).

This concurs with hydropower likewise being considered, among other things, central to achieving developmental objectives, including in the Mekong Region where national development and modernisation narratives emphasise hydropower expansion (see Rousseau and Habich-Sobiegalla, 2021). The sediments trapped behind dams thus emerge as political objects, to the same extent as the sediments poured into concrete-built infrastructure and those that travel all the way to the South China Sea.

Hydropolitics scholarship: Placing sand in regional water politics

The Mekong Region is the focus of an ample body of scholarship analysing series of interlinked water–society relations. A significant part of this literature situates itself within research fields spanning broadly defined research agendas such as hydropolitics, political ecology, or political economy (recent reviews include Middleton (2022) and Rogers et al. (2023)). Recurrent themes in this literature have long included the political processes and institutions that participate in water governance, and the transboundary implications of hydraulic infrastructure development (e.g., Hirsch, 2020; Rogers et al., 2023). Hydropower expansion has been a key focus, including dam development along the Lancang River (upper Mekong) and further downstream along both the Mekong and tributary rivers (e.g., Soukhaphon et al., 2021). In contrast, the sediments themselves are seldom framed as a key transboundary resource that deserves to be shared, governed, and managed in the same way as water; sand, rather, goes 'unseen' (key exceptions include Grundy-Warr and Lin (2020), Hirsch (2020), and Sopera (2022).

While the cumulative impact of various activities on the Mekong sediment budget is relatively well addressed in environmental science (cf. Hecht, 2019; Jordan et al., 2019; Le et al., 2020; Schmitt et al., 2017), social science research seldom approaches the many drivers of environmental change at once in this region (exceptions include Baird and Barney (2017) and Rousseau and Marschke (2023)). Baird and Barney (2017: 788) explain that this incomplete picture "facilitates managerialist legibility, at the expense of peasant resource users" who experience real livelihood challenges. The same authors argue that political will is required to overcome such a de-politicised and siloed approach, and that the Mekong Region has so far experienced a deficit in that regard. We probe whether this conclusion applies to how the MRC has addressed the interactions between hydropower and sand mining, building from institutional discourse analysis, an approach common to scholarship centred on the broadly defined hydropolitics of the Mekong Region.

The case for discourse analysis

Discourses explicitly or implicitly communicate the values and meanings that specific social actors uphold and/or promote. These significances emerge within various social practices, including communication (oral, visual, written, sensory, etc.), practices, and policies that testify to how actors approach knowledge, truth, and power (della Faille, 2011; Gerlak and Schmeier 2014). 'Discursive' means including emphasis or omissions and can materialise in rules, strategies, and semantic categories, among others.

In the Mekong Region, this type of analysis has been frequently utilised to decipher the discourses, narratives, biases, and agendas that institutions, including the MRC, uphold towards planned hydropower schemes (Bakker, 1999; Fox and Sneddon, 2019; Geheb et al., 2015; Middleton, 2022). These analyses—and many others—have shed light on the knowledge production strategies utilised to legitimise dam schemes, and to, conversely, pre-empt dissenting counter-narratives challenging official discourses and the authority of the individuals and regimes perpetuating them. Retracing institutional discourses also allows the evolution of institutional analyses to be accounted for (Grünwald et al., 2022). Besides discourses centred on hydropower, scholars have also probed the MRC's discourses on climate change (Gerlak and Schmeier, 2014), and security (Gerlak and Mukhtarov, 2016).

That being said, to our knowledge, regional discourses on changing sediment budgets and their drivers have not been the focus of earlier inquiry. Given the socio-environmental importance of sand and the growing consequences of its extraction in the Mekong Region, we argue that our contribution fills an important gap. The MRC documents that we build from highlight political narratives that are shaped by the member states' priorities, compromises, and omissions, and this chapter deciphers some of the circumstances at play in evolving MRC sand narratives.

Context: The Mekong's sand

The Mekong River's 'natural' sediment discharge was said to amount to 160 million tonnes in the 1980s (Milliman and Syvitski, 1992) and half of it emerged from the Chinese upstream (Kummu and Varis, 2007),[4] while the final 1,000 km section of the river is where most accretion occurs (Gupta and Liew, 2007). As dams were erected in the Mekong watershed, so was sediment increasingly trapped by reservoirs. By the early 2000s, it was estimated that 35-45 tonnes of sediment were trapped yearly as a result of upstream damming, and estimates suggest that this value could reach 100 million tonnes if all dams currently planned in the watershed as a whole are built (Jordan et al., 2019). Sand mining also takes an important toll on the Mekong sediment discharge. Many sand mining activities remain unregulated, and sediment withdrawal estimates mostly rest on GIS analysis, including sand barge counting (i.e., Gruel et al., 2022; Hackney et al., 2021; Kumar et al., 2024). Some evaluations suggest that anywhere between 7.7 and 20 million tonnes of aggregates are dredged annually in the Vietnamese delta region and 59 million tonnes yearly in Cambodia alone (Hackney et al., 2021; Jordan et al., 2019), driven in great part by the massive

[4] Kummu and Varis (2007) argue that currents downstream of Kratie make it difficult to estimate the total sediment discharge for the whole watershed.

land and marsh infilling programme that has unfolded in and around Phnom Penh over the last decades (Cyan Bird, 2020; Marschke et al., 2022). No exhaustive value is available for sand extraction in the Mekong Basin as a whole (Gruel et al., 2022).

What is clear though is that the combined impacts from hydropower expansion and sand mining have accrued rapidly over the last few decades. Mainstream and tributary river damming led to massive reductions in sediment discharge values between the 1980s (pre-damming) and mid-2010s, with reductions in sediment budgets ranging from about 75 per cent in the Vietnam Mekong Delta to up to 90 per cent in Northern Thailand (Park et al., 2022). Current estimates forecast that the future amount of sediment discharging into the South China Sea may total just 5 per cent of the 'natural' value (Jordan et al., 2019). Reductions in the sediment load drive commensurate impacts on fisheries and agriculture (e.g., Intralawan et al., 2019) which remain core livelihood activities and central to the food security of the Mekong Region. Reduced flooding has meant that sediment deposition affects agriculture, in turn leading to increased erosion, reduced fertility, and seawater intrusion from the Vietnam Mekong Delta—itself aggravated by sea-level rise—leading to soil or irrigation water salinisation (Intralawan et al., 2019; Park et al., 2022). We now turn to the methodology which we developed to probe the MRC's discourse with regard to such sand challenges.

Methodology

We examined how sand was approached in MRC reports publicly available from the organisation's English language web portal by the time we carried out data collection for this research in July 2023. Our analysis concentrated on reports classified under four categories on the MRC web portal 'publications' page, namely the 'Basin Reports' (n = 57), 'Management and Development Series' (n = 23), 'Policies, Procedures and Guidelines' (n = 32), and the 'MRC Strategic Plan' (n = 5) (Fig. 7.1). The latter plans are the MRC's flagship publications, and we aligned the time periods we emphasise in our results with each strategic plan found in our study: 2001-2005 (n = 15), 2006-2010 (n = 22), 2011-2015 (n = 31), 2016-2020 (n = 36), and 2021-July 2023 (n = 13) (Fig. 7.2).[5]

[5] We do not intend for this selection to be exhaustive, although it provides a clear snapshot of how the MRC presented itself on the internet when we carried out our search. The organisation has been undergoing massive website updates over the last few years and publicly available publications along with how they are categorised have been shifting since. Key documents absent from our database are those from the 2017-2018 'Study on the Sustainable Management and Development of the Mekong River Basin including impacts of Mainstream Hydropower Projects', also known as the Council Study.

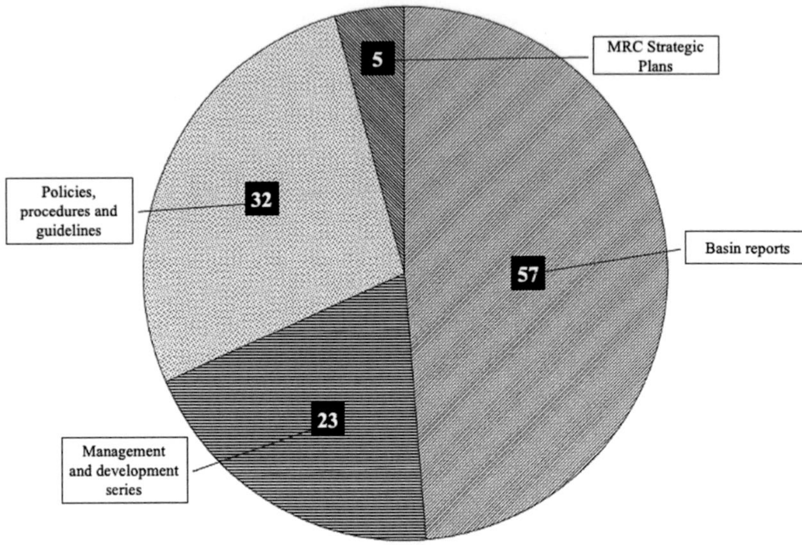

Figure 7.1: Types of reports included in our sample.

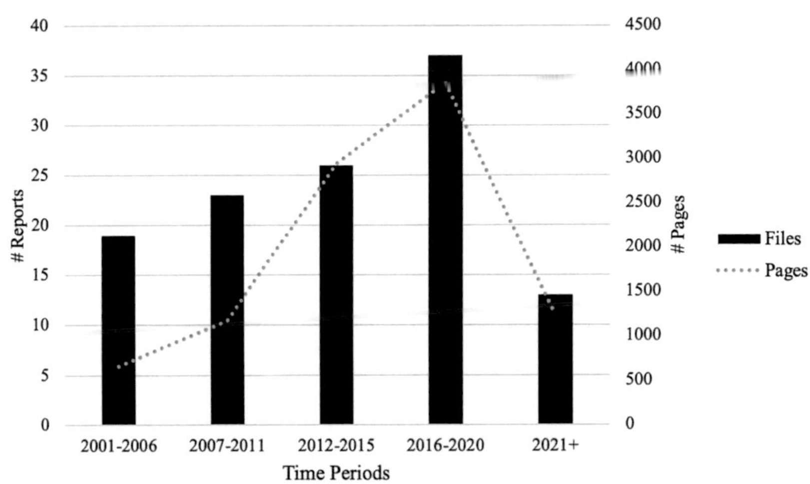

Figure 7.2: Number of sampled MRC reports and total number of pages per time period.

Our strategy yielded a corpus that was simultaneously sufficient to account for the evolution of MRC narratives over time, while still being manageable. In sum, this approach led us to probe 117 English-language documents totalling 9,690 pages. This amounts to only a fraction of the massive literature generated at the MRC since its inception in 1995. Still, our sample included widely diffused reports from the Commission and offers an exhaustive view of the sand discourses that the MRC has emphasised through the years.

The selected documents were imported to NVivo (v.1.7.1), a widely used computer-assisted qualitative data analysis application (CAQDAS), which we utilised to conduct text searches, determine code frequencies, and capture the corresponding paragraphs for context. We developed a three-step coding strategy that involved a priori, axial, and thematic coding.

We first established, through a priori coding, a series of themes related to our aim and questions emerging from our inquiry, namely:

- How do MRC documents approach the impacts of hydropower development and those of sand mining on the Mekong river's sediment budget?
- Are the cumulative impacts from these individual activities addressed?
- Are sediments framed as a transnational resource or not?

Hence, we aimed to identify information on: (a) sand; (b) hydropower; and (c) resource management and organised the a priori codes into these three categories.

In the axial coding phase, our methodology involved refining our preliminary set of codes to better align with the MRC's perspectives on sediment changes and their drivers. During this refinement process, several codes that did not fit the context or were redundant were removed. For instance, while 'erosion' is known to be a consequence of sand mining, MRC documents mainly discuss erosion in relation to other causes or activities. Likewise, while we first expected 'sediment' to connect with sand-/sand mining-related results, this notion is almost exclusively addressed in hydropower-related discussions in the sampled reports. The refined list of codes was used in compound searches to identify co-occurrence of terms, and matrix coding queries were developed to visualise these co-occurrences.

Additionally, we examined the broad themes above over the quinquennial intervals. Utilising a Python script which extracts and renames documents based on their XML metadata, we sorted each document into its corresponding temporal period. Our aim was to discern any trends or shifts in the MRC's emphasis on the treatment of sand throughout these intervals. Given the varied distribution of report lengths across these periods, we adopted a normalisation technique as outlined by Neuendorf (2017). Specifically, we computed code occurrences per 100 pages to ensure consistency and comparability across temporal periods.

Quantifying sand occurrences

Sand-related codes appeared much less frequently in the sampled reports than those related to hydropower and resource management (Table 7.1). This is partly related to sand-related codes being more specific than those related to resource management. The more general codes appearing much more frequently also reflects the broad nature of the MRC mandate.

Table 7.1: Themes, codes, and occurrences in the sampled documents.

Sand	Frequency	Hydropower	Frequency	Resource management	Frequency
Aggregate	114	Dam	2,992	Development	6,121
Sand	389	Hydroelectricity	17	Governance	249
Sand dredging	13	Hydropower	4,183	Management	4,795
Sand mining	74	Reservoir	1,596	Resource	3,238
Sediment extraction	24	Sediment	2,221	Resource governance	6
				Resource management	144
Total	614	Total	11009	Total	14,553

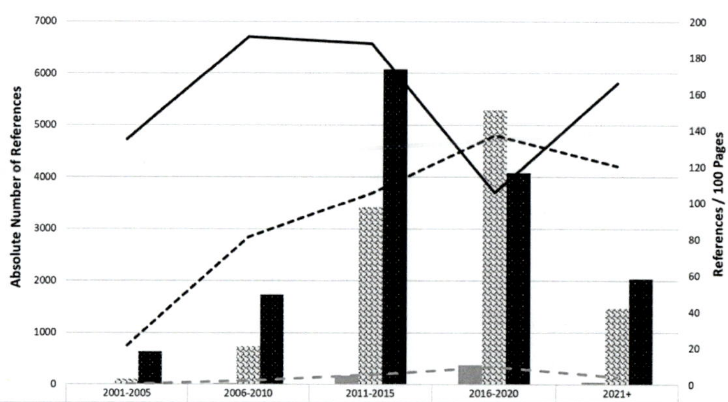

	2001-2005	2006-2010	2011-2015	2016-2020	2021+
Sand	0	18	174	373	49
Sand (Per 100 Pages)	0	1.99	5.38	9.68	4
Hydropower	101	734	3409	5289	1476
Hydropower (Per 100 Pages)	21.49	81.19	105.35	137.23	120.39
Resources (RM)	634	1731	6074	4080	2034
RM (Per 100 Pages)	134.89	191.48	187.7	105.86	165.91

Figure 7.3: Temporal occurrence of themes.

'Sustainable resource management' is at the core of the MRC mandate, and related codes remained central within MRC documents throughout the period of focus (Fig. 7.3). Hydropower is also a key theme that constantly appears in the reports consulted. While upstream hydropower developments along the Lancang River have remained an important focus throughout the time interval we probed, projects planned or unfolding along the lower section of the watershed have received increasing attention in MRC reports from the early 2010s onwards.

Sand contrasts sharply with the other two themes. Sand is a relative latecomer in MRC narratives, and related codes appear much less frequently in the sampled reports than those that connect with the other two themes. This trend concurs with how sand has been approached in other fields, including in academic writing and policy documents, with sand really emerging as an object of inquiry from the late 2010s onward, in the wake of the first UNEP Sand and Sustainability Report (UNEP, 2019; see also Suykens et al., 2023).

Looking in greater depth into how the MRC approaches sand and how it relates to our other focal themes, quantitative results from axial coding procedures highlight sand as most often being related to hydropower related codes, beginning with sediments (Table 7.2). Little connection is made between sand and resource management codes, as sound sand management is only a recent and emerging agenda in MRC documents, which increasingly acknowledge the interconnections between sand and development.

Table 7.2: Axial coding results: Co-occurrence of sand codes vis-à-vis other codes.

	Aggregate	**Sand**	**Sand dredging**	**Sand mining**	**Sediment extraction**	**Row Total**
Dam	6	51	0	16	7	80
Hydroelectricity	0	0	0	0	0	0
Hydropower	6	21	0	19	8	54
Reservoir	4	31	0	8	0	43
Sediment	6	122	1	25	0	154
Development	9	28	2	15	7	61
Governance	0	1	0	0	0	1
Management	1	22	0	7	5	35
Resource	1	7	1	8	0	17
Resource governance	0	0	0	0	0	0
Resource management	0	0	0	0	0	0
Column Total	33	283	4	98	27	445

Evolving sand discourses

As alluded to when discussing our quantitative results, sand does not occupy an important place in MRC narratives, and sand mining and its ecological impacts in the Mekong Region only recently emerged among the Commission's concerns. Reports have highlighted that decreased sediment budgets lead to impacts spanning increased erosion; habitat loss (including sand bar destruction), riverbed incision, reduced connectivity to floodplains, reduced flood plain productivity (including as a consequence of increasing salt intrusion), and biodiversity reduction (MRC, 2015b, 2019a, 2019b, 2020, 2021).

Current and potential impacts from sand mining are considered most acute in the Mekong Delta, one reason for which is that suspended sediment in this area is finer than in other sections of the river (MRC, 2015b). Whereas sand is transported both as suspended sediments and bedload upstream, it only travels as bedload in the delta, due to the lower slope. This flatter topography also means that few sediments are generated within the delta per se, with most sediments originating from upstream areas or from coastal reshuffling (MRC, 2015b). Land subsidence—itself amplified by groundwater extraction—and saline water intrusion also manifest acutely in the delta, adding to the circumstances that make this zone particularly sensitive to sediment budget changes. According to the MRC, reduced sand replenishment and its impacts in the delta require 'urgent' attention in a context of increased development activities and sediment extraction (MRC, 2021: 2).

The sampled reports frequently lament the lack of quantified data on sediment budget changes and associated impacts, a concern echoed elsewhere (Bendixen et al., 2019; Katz-Lavigne, Pandey and Suykens, 2022). The MRC acknowledges many unknowns pertaining to sand and extraction activities (MRC, 2019b), including the Mekong River's total sediment budget, its historical evolution and the extent to which the sediment budget is composed of clay, silt, sand, and/or coarser aggregates (MRC 2015b).

Available information states that the total annual sediment budget at Pakse was estimated at 160 mt yearly prior to mainstream hydropower development, with about 60 per cent of this volume originating from the Lancang River, 10 per cent from the Sesan, Srepok, and Sekong Rivers (the 3S system), and the remaining 30 per cent from other tributary rivers (MRC, 2015b). Sediments originating from the upstream section of the watershed are coarser than those generated downstream, with the former mainly composed of sand, and silt comprising most of the latter downstream content (MRC, 2015b). One estimate suggests that the sediment load recorded at Pakse in 2015 was 55 per cent lower than historical values due to the combined impacts from reservoir trapping and sand mining (MRC, 2019b). Retroaction between the consequences of both reservoir trapping and sand mining is acknowledged, including that sediment extraction amplifies the impacts of upstream reservoir trapping on erosion. Increasing erosion in downstream areas in turn attenuates total sediment budget changes in the basin as a whole (MRC, 2020). Associated changes in the characteristics of sediments, including diameter and transport, remain unknown, however.

The most widely stated data on sediment extraction rates date from a 2013 study (Bravard et al., 2013) estimating that 35 million m^3 of sand (equivalent to 55 million tonnes) were extracted from the Mekong River in 2011.[6] In that same year, it is believed that 60 per cent of extraction occurred in Cambodia alone (MRC, 2019b). There is uncertainty about the economic value of extracted aggregates. USD 3 per m^3 is mentioned as the market value of sand and gravel (MRC, 2015a). Accordingly, sand mining was 'believed' to have generated USD 175 million yearly in the Mekong in the late 2010s (MRC, 2019b: ix). That said, reports highlight that sand mining remains an opaque sector, partly because of a lack of resources to monitor sand extraction and enforcing legal and regulatory guidelines (MRC, 2012, 2019b). Hence, while doubts persist about the accuracy of published figures, recent extensive infill and infrastructure development trends suggest that they have grown rapidly and massively.

No exhaustive figure is available on the value of the ecological services that sand sustains (MRC, 2019b). One report mentions that riverbank erosion in Laos leads to the disappearance of land valued at USD 100,000 yearly, but no further details are provided (MRC, 2019b). Little is known about the importance of sand mining-driven consequences on other economic sectors, including agriculture and fisheries (MRC, 2019b). Still, it is foreseen that changes in sediment transport, including grain sizes, the distances over which they travel and the seasonality of these movements affect the availability of the nutrients fundamental to these sectors (MRC, 2019a), themselves key to local economies and livelihoods. Sand mining also fosters a whole series of downstream sectors, including housing and infrastructure development, but the extent to which sand sustains economic activities beyond these sectors is not well understood (MRC, 2019b). In any case, the MRC acknowledges that understanding the value of sand from a financial angle alone is insufficient (MRC, 2020).

Sand and sand mining's recent emergence in MRC narratives stems from the fact that not much research—including longitudinal inquiry—has been carried out in the Mekong Region so far (cf. MRC, 2011). The MRC acknowledges that this gap

[6] While the analysis of the Council Study goes beyond the scope of this chapter for reasons which we explained above, it is noteworthy to highlight that the Study does not introduce further data about sand mining activities, on the grounds that no such data are available (MRC 2017). Still, the study notes that "sediment mining in the Mekong River channel has been a politically sensitive issue" (MRC 2017: 12)—which offers one possible explanation for this data gap that we have not encountered elsewhere in MRC documents. Another noteworthy distinction between the Council Study and our dataset is that the former approaches dredging operations carried out in the course of navigation channel improvement (an activity about which only 'limited data is available') distinct from sand mining operations.

must be addressed, including through updating monitoring guidelines so that they further account for the increasing influence of sand mining on the river's sediment budget (MRC, 2011). Information exchange mechanisms between riparian countries (including upstream countries) concentrate on water discharge and should be updated (MRC, 2019b). Also at play are the challenges to isolate the impacts of sand mining alone from those that other anthropogenic activities and natural processes drive. Besides reservoir trapping, other circumstances involved include land use practices (including land clearing and infill), climate change, agricultural practices (including irrigation), infrastructure development, riverbank armouring, etc. (MRC, 2014a; 2015b; 2016; 2020). All of these factors are typically approached generally, with no specific project flagged as detrimental to the sediment budget nor the riparian livelihoods that depend on it. For instance, infill schemes in Cambodia and their spectacular sand requirements are not explicitly mentioned. With regard to sediment trapping, specific attention is placed on large upstream dams located along the Lancang River, whereas the influence from run-of-river and/or tributary dams is acknowledged in a more general fashion (MRC, 2019b). That said, sand mining is considered as an activity that can exacerbate the consequences of other change drivers (MRC, 2015b), and the MRC anticipates that anthropogenic drivers are overall set to increasingly exert greater influence on sediment budgets compared to geomorphological processes (MRC, 2011).

While MRC reports published in earlier periods selectively stress the proximal causes from sediment budget changes, a shift in discourses began in the mid-2010s, with ultimate causes being increasingly considered. Hence, broader processes such as industrialisation, urbanisation, and deforestation are now listed among factors driving change in sediment budgets (MRC, 2016), but in general terms only. Discourses on the relationship between hydropower and sand mining are also evolving. For instance, while reports published earlier solely present sand mining and hydropower reservoir trapping as two activities driving accruing impacts on the Mekong's sediment budget, questions on the implications of reservoir trapping on the sand mining sector are now being raised (MRC, 2015b). This echoes press reports on the challenges of securing aggregate stocks necessary to develop transport infrastructure in Vietnam, including in the Vietnam Mekong Delta and beyond (Tran, 2023; Vo, 2023; Xuan, 2023). One MRC report goes as far as to consider the widespread availability of cheap aggregates as a thing of the past, and claims that changes in development pathways are now required (MRC, 2020).

Recommendations to alleviate sand scarcity and the related impacts include utilising non-conventional concrete products with a lower aggregate content (MRC, 2014b). Other policy recommendations to achieve sounder sand management span establishing new information exchange mechanisms with upstream countries, so that hydrologic and reservoir operation data become more widely available (MRC,

2019b). Sediment monitoring must be enhanced, including through the use of remote sensing technologies (MRC, 2021). Great attention is placed on improving sediment management practices in existing reservoirs and optimising sediment sluicing and flushing both at the design and operation stages so that coarser aggregates—rather than silt only—are also released from reservoirs (MRC, 2019a; 2020). That said, such practices come with a price, which could amount to USD 51 million yearly and represent up to 2.8 per cent of the total revenue from the hydropower sector in the region (MRC, 2020). For reservoirs with no sediment flushing capacities, dredging sediments and transporting these to downstream areas are considered, although the economic feasibility of such an approach is questioned (MRC, 2020).

The MRC also calls for the elaboration of sustainable sand mining guidelines (MRC, 2020). These guidelines could encompass a ban on sand mining downstream of the mainstream Mekong dams, where little to no natural sediment replenishment occurs. The main channel from the Mekong, Bassac, and Tonle Sap Rivers is considered especially significant to ensuring channel stability in the Mekong Delta (MRC, 2020). Planned hydropower projects along with the guidelines shaping future projects should, the MRC argues, be revised to ensure that they do not further impact sediment transport (MRC, 2020). Other mitigation strategies include the promotion of reservoir dredging, and changing the perceptions of sand. Currently considered as a cheap land reclamation and construction commodity, sand needs to be framed as a valuable environmental asset (MRC, 2020). Research capacity must also be enhanced and new research technologies deployed further, including grain-size analysers and remote sensing (MRC, 2020).

Conclusion: A (too?) ambitious agenda?

The data and analysis above highlight that sand is emerging as a key theme in MRC narratives. The recent intensification of extraction, along with accruing consequences, appear more commonly in MRC documents, and are approached in an increasingly nuanced manner. That said, many information gaps remain, partly because it is only recently that sand has emerged as 'strategic'. As the MRC acknowledges, achieving better sediment management is fraught with challenges, including, on the one hand, those that pertain to enhancing information-sharing mechanisms between member countries and their upstream neighbours and, on the other hand, those related to implementing national regulatory guidelines at the local scale. That said, also likely at play are the secretive nature of the sand mining industry and the corruption that often surrounds it. Additional complications pertain to dominant urbanisation and development models as well the strategic importance allocated to the individual infrastructure and infill schemes that fuel regional sand demand.

The combined impacts of sand mining and hydropower expansion on sediment budgets are acknowledged, as highlighted in both our quantitative data and the narratives that we probed. Also important is that this relation is increasingly approached in a manner that transcends the individual impacts of sediment trapping and sand mining on the Mekong's sediment budget. Still, many uncertainties remain about these individual contributions and how their interactions shape specific impacts. The influence of reservoir trapping on downstream extraction activities is considered, but the very fact that hydropower dam building also requires massive amounts of sand is not raised whatsoever in the hydropower-coded result. The same goes for the sand requirements from other individual economic sectors or types of infrastructure, including, ironically, those intended to protect infrastructure and riverbanks from erosion.

Nonetheless, MRC narratives on sand are increasingly ambitious. The MRC acknowledges that sand management transcends the national scale and involves transboundary responsibilities, including with regard to reservoir management and sediment extraction coordination (MRC, 2021). The Commission considers the need to reevaluate sand resources vis-à-vis the economic benefits derived from their extraction as a priority in a context where the impacts from climate change are likely to accrue in the short term (MRC, 2021). It is now acknowledged that "continuing uncoordinated development" (MRC, 2021: 52) renders the search for solutions more complicated and costly, and that combined anthropogenic pressure might lead to a 'step-change' scenario (MRC, 2021: 40). In lieu, the MRC calls for achieving an "acceptable balance between ecological function, economic development and climate resilience" (MRC, 2021: 3).

Whether the current outlook involves such an acceptable balance is open to debate. Indeed, we note the gap between the technocratic and de-politicised documents which we reviewed and the emergency that riparian communities and authorities express when we probe sand budget changes in the Mekong Region and beyond. The distress that riverside households experience when their houses and land fall into the river has been increasingly documented via local, national, and international media outlets, as is riverside infrastructure failure. These circumstances yield previously unknown accountability challenges for local and national authorities alike. Changing sediment budgets also pose threats to the very territorial integrity of the Mekong countries—not least the current and continuing reductions to the Vietnamese Mekong Delta area. Our data do not allow us to conclude that these new circumstances explain why MRC discourses on sand have evolved, but the emerging sand narratives are welcome in a context where less ambitious structural changes, such as adopting less sand-intensive concrete materials, will likely not generate sufficient sand demand reductions to mitigate rapidly changing environmental circumstances (cf. Bisht, 2022). Still, beyond the broad and general critiques about regional development models that the MRC now relays, questions could also be raised towards individuals and specific infrastructure or infill schemes. This is an area into which the MRC documents that we probed do not venture.

The MRC's ability to convince member states and upstream nations to implement the solutions and development models that it increasingly calls for will be crucial to the region's sand future. Unfortunately, water-centred research on hydropolitics in the Mekong has long raised doubts about the Commission's capacity to ensure that the structural changes that it promotes are implemented amongst its member countries. Convincing members to adopt the sound sand management policies that it advocates for is likely to face similar inertial force, as it would involve revisiting the infrastructure-led development model that is hegemonic in the region (and beyond) and deprive powerful social actors of the benefits that they gain from it. In any case, we claim that sand needs to be added to the resources that this hydropolitics scholarship focuses on, and we hope that the increasing attention that the academic and governance communities allocate to this resource will allow the barriers to its sounder governance to be mitigated.

References

Bakker, K. 1999. The politics of hydropower: developing the Mekong. *Political Geography* 18(2): 209–232, https://doi.org/10.1016/S0962-6298(98)00085-7.

Baird, I.G. and Barney, K. 2017. The political ecology of cross-sectoral cumulative impacts: modern landscapes, large hydropower dams and industrial tree plantations in Laos and Cambodia. *The Journal of Peasant Studies* 44(4): 769–795, https://doi.org/10.1080/03066150.2017.1289921.

Beiser, V. 2019 *The world in a grain: the story of sand and how it transformed civilization*. London: Penguin.

Bendixen, M., Best, J., Hackney, C. and Iversen, L.L. 2019. Time is running out for sand. *Nature* 571(7763): 29–31, https://www.nature.com/articles/d41586-019-02042-4.

Bikkina, N. and Tejo Shravya, K.V. 2017. Women's self-help groups in the sand market. *International Journal of Rural Management* 13(1): 108–114, https://doi.org/10.1177/0973005217693338.

Bisht, A. 2021. Conceptualizing sand extractivism: deconstructing an emerging resource frontier. *The Extractive Industries and Society* 8(2): 100904, https://doi.org/10.1016/j.exis.2021.100904.

Bisht, A. and Gerber, J.F. 2017. Ecological distribution conflicts (EDCs) over mineral extraction in India: an overview. *The Extractive Industries and Society* 4(3): 548–563, https://doi.org/10.1016/j.exis.2017.03.008.

Bravard, J.-P., Goichot, M. and Gaillot, S. 2013. Geography of sand and gravel mining in the lower Mekong River: first survey and impact assessment. *EchoGéo* 26 (2013): 1–20, https://doi.org/10.4000/echogeo.13659.

Cyan Bird, K.D. 2020. *Quantifying the infilling of Mekong floodplains in Cambodia using DEM differencing*. M.Sc. thesis. Exeter: University of Exeter, https://ore.exeter.ac.uk/articles/thesis/Quantifying_the_

Infilling_of_Mekong_Floodplains_in_Cambodia_using_DEM_Differencing/29777681?file=56809502 (accessed 7 January 2026).

Da, S. and Le Billon, P. 2022. Sand mining: Stopping the grind of unregulated supply chains. *The Extractive Industries and Society* 10(2022): 101070, https://doi.org/10.1016/j.exis.2022.101070.

della Faille, D. 2011. Discourse analysis in international development studies: mapping some contemporary contributions. *Journal of Multicultural Discourses* 6(3): 215–235, https://doi.org/10.1080/17447143.2011.594512.

Fox, C. and Sneddon, C. 2019. Political borders, epistemological boundaries, and contested knowledges: constructing dams and narratives in the Mekong River Basin. *Water* 11(2019): 413, https://doi.org/10.3390/w11030413.

Franks, D.M. 2020. Reclaiming the neglected minerals of development. *The Extractive Industries and Society* 7(2): 453–460, https://doi.org/10.1016/j.exis.2020.02.002.

Geheb, K., West, N. and Matthews, N. 2015. The invisible dam: hydropower and its narration in the Lao Peoples' Democratic Republic. In Matthews, N. and Geheb, K. (eds.), *Hydropower development in the Mekong Region: political, socio-economic and environmental perspectives*, 101–126. Abingdon: Routledge.

Gerlak, A.K. and Schmeier, S. 2014. Climate change and transboundary waters: a study of discourse in the Mekong River Commission. *The Journal of Environment and Development* 23(3): 358–386, https://doi.org/10.1177/1070496514537276.

Gerlak, A.K. and Mukhtarov, F. 2016. Many faces of security: discursive framing in cross-border natural resource governance in the Mekong River Commission. *Globalizations* 13(6): 719–740, https://doi.org/10.1080/14747731.2015.1134133.

Gruel, C.-R., Park, E., Switzer, A.D., Kumar, S., Loc Ho, H., Kantoush, S., Van Binh, D. and Feng, L. 2022. New systematically measured sand mining budget for the Mekong Delta reveals rising trends and significant volume underestimations. *International Journal of Applied Earth Observation and Geoinformation* 108(2022): 102736, https://dx.doi.org/10.1016/j.jag.2022.102736.

Grundy-Warr, C. and Lin, S. 2020. The unseen transboundary commons that matter for Cambodia's inland fisheries: changing sediment flows in the Mekong hydrological flood pulse. *Asia Pacific Viewpoint* 61(2): 249–265, https://doi.org/10.1111/apv.12266.

Grünwald, R., Wang, W. and Feng, Y. 2022. Politicization of the hydropower dams in the Lancang-Mekong Basin: a review of contemporary environmental challenges. *Energies* 15(5): 1682, https://doi.org/10.3390/en15051682.

Gupta, A. and Liew, S.C. 2007. The Mekong from satellite imagery: a quick look at a large river. *Geomorphology* 85(3): 259–274, https://doi.org/10.1016/j.geomorph.2006.03.036.

Hackney, C.R., Vasilopoulos, G., Heng, S., Darbari, V., Walker, S. and Parsons, D.R. 2021. Sand mining far outpaces natural supply in a large alluvial river. *Earth Surf. Dynam.* 9(5): 1323–1334, https://doi.org/10.5194/esurf-9-1323-2021.

Hecht, J.S., Lacombe, G., Arias, M.E., Dang, T.D. and Piman, T. 2019. Hydropower dams of the Mekong River basin: A review of their hydrological impacts. *Journal of Hydrology* 568: 285–300, https://doi.org/10.1016/j.jhydrol.2018.10.045.

Hirsch, P. 2020. Scaling the environmental commons: broadening our frame of reference for transboundary governance in Southeast Asia. *Asia Pacific Viewpoint* 61(2): 190–202, https://doi.org/10.1111/apv.12253.

Intralawan, A., Smajgl, A., McConnell, W., Ahlquist, D.B., Ward, J. and Kramer, D.B. 2019. Reviewing benefits and costs of hydropower development evidence from the Lower Mekong River Basin. *WIREs Water* 6(4): e1347, https://doi.org/10.1002/wat2.1347.

Jordan, C., Tiede, J., Lojek, O., Visscher, J., Apel, H., Nguyen, H.Q., Quang, C.N.X. and Schlurmann, T. 2019. Sand mining in the Mekong Delta revisited – current scales of local sediment deficits. *Scientific Reports* 9(1): 17823, https://doi.org/10.1038/s41598-019-53804-z.

Katz-Lavigne, S., Pandey, S. and Suykens, B. 2022. Mapping global sand: extraction, research and policy options. GiC Network: https://www.cmi.no/publications/8586-mapping-global-sand-extraction-research-and-policy-options (accessed 7 January 2026).

Koehnken, L., Rintoul, M.S., Goichot, M., Tickner, D., Loftus, A.-C. and Acreman, M.C. 2020. Impacts of riverine sand mining on freshwater ecosystems: a review of the scientific evidence and guidance for future research. *River Research and Applications* 36(3): 362–370, https://doi.org/10.1002/rra.3586.

Kumar, S., Park, E., Tran, D.D., Wang, J., Loc Ho, H., Feng, L., Kantoush, S.A., Binh, D.V., Dongfeng, L. and Switzer, A.D. 2024. A deep learning framework to map riverbed sand mining budgets in large tropical deltas. *GIScience & Remote Sensing* 61(1): 2285178, https://doi.org/10.1080/15481603.2023.2285178.

Kummu, M. and Varis, O. 2007. Sediment-related impacts due to upstream reservoir trapping, the Lower Mekong River. *Geomorphology* 85(2007): 275–293, https://doi.org/10.1016/j.geomorph.2006.03.024.

Lamb, V., Marschke, M. and Rigg, J. 2019. Trading sand, undermining lives: omitted livelihoods in the global trade in sand. *Annals of the American Association of Geographers* 109(5): 1511–1528, https://doi.org/10.1080/24694452.2018.1541401.

Langer, W. 2016. Sustainability of aggregates in construction. In Khatib, J.M. (ed.), *Sustainability of Construction Materials* (Second Edition), 181–207. Sawston: Woodhead Publishing.

Le, H.-A., Gratiot, N., Santini, W., Ribolzi, O., Tran, D., Meriaux, X., Deleersnijder, E. and Soares-Frazão, S. 2020. Suspended sediment properties in the Lower Mekong River, from fluvial to estuarine environments. *Estuarine, Coastal and Shelf Science* 233(2020): 106522, https://doi.org/10.1016/j.ecss.2019.106522.

Li, D., Lu, X., Walling, D.E., Zhang, T., Steiner, J.F., Wasson, R.J., Harrison, S., Nepal, S., Nie, Y., Immerzeel, W.W., Shugar, D.H., Koppes, M., Lane, S., Zeng, Z., Sun, X., Yegorov, A. and Bolch, T. 2022. High mountain Asia hydropower systems threatened by climate-driven landscape instability. *Nature geoscience* 15(7): 520–530, https://doi.org/10.1038/s41561-022-00953-y.

Marschke, M. Rousseau, J.-F., Beckwith, L. and van Arragon, L. 2022. Displaced sand, displaced people: the livelihood impacts of sand mining. In Linh, H.T.P., Lagrée, S., Espagne, É. and Drogoul, A. (eds.), *Inequalities and environmental changes in the Mekong Region*, 85–104. Paris: AFD éditions.

Marschke, M. and Rousseau, J.-F. 2022a. Unpacking Cambodia's sand industry: livelihoods, ecologies and development. Köln: AG Resourcen/Asienhaus: https://www.asienhaus.de/archiv/asienhaus/Ressourcenkonflikte/Sand/SAH_2022_Sand_Cambodia_Marschke_Rousseau_EN.pdf (accessed 7 January 2026).

Marschke, M. and Rousseau, J.-F. 2022b. Sand ecologies, livelihoods and governance in Asia: a systematic scoping review. *Resources Policy* 77(2022): 102671, https://doi.org/10.1016/j.resourpol.2022.102671.

Mercado, L. and Goichot, M. 2023. Time to change the myopic view of the Mekong. *Asia Times*. 4 April 2023: https://asiatimes.com/2023/04/time-to-change-myopic-view-of-the-mekong/ (accessed 7 January 2026).

Milliman, J.D. and Syvitski, J.P.M. 1992. Geomorphic/Tectonic Control of Sediment Discharge to the Ocean: The Importance of Small Mountainous Rivers. *The Journal of Geology* 100(5): 525–544, https://doi.org/10.1086/629606.

MRC (Mekong River Commission). 2011. *Assessment of basin-wide development scenarios (Main Report)*. Vientiane, Lao PDR: Mekong River Commission Secretariat, https://doi:10.52107/mrc.ajhykw.

MRC (Mekong River Commission). 2012. Carriage, handling and storage of dangerous goods along the Mekong River: risk analysis (Volume I). Vientiane: MRC Secretariat, https://doi:10.52107/mrc.ajgt3g.

MRC (Mekong River Commission). 2014a. Guiding considerations on transboundary monitoring for LMB hydropower planning and management. Vientiane: MRC Secretariat, https://doi:10.52107/mrc.ajg8a8.

MRC (Mekong River Commission). 2014b. Technical capacity review of the irrigation sector. Vientiane: MRC Secretariat, https://doi:10.52107/mrc.ajgfc6.

MRC (Mekong River Commission). 2015a. Guidelines for the evaluation of hydropower and multi-purpose project portfolios: economics practice guide (Annex 1): economics practice guide. Vientiane: MRC Secretariat, https://doi:10.52107/mrc.ajg81x.

MRC (Mekong River Commission). 2015b. Hydropower risks and impact mitigation manual (Volume 2, 1st Interim Report). Vientiane: MRC Secretariat, https://doi:10.52107/mrc.ajg81k.

MRC (Mekong River Commission). 2016. Strategic plan 2016-2020. Vientiane: MRC Secretariat, https://doi:10.52107/mrc.ajg7tk.

MRC (Mekong River Commission). 2017. The Council Study: report for navigation thematic area Cambodia – Lao PDR – Thailand – Viet Nam. Vientiane: MRC Secretariat.

MRC (Mekong River Commission). 2019a. Hydropower mitigation guidelines (Volume 2). Vientiane, Lao PDR: Mekong River Commission Secretariat, https://doi:10.52107/mrc.ajg6cq.

MRC (Mekong River Commission). 2019b. State of the basin report 2018. Vientiane: MRC Secretariat, https://doi:10.52107/mrc.ajg54f.

MRC (Mekong River Commission). 2020. Hydropower mitigation guidelines (Volume 3). Vientiane, Lao PDR: MRC Secretariat, https://doi:10.52107/mrc.ajgo06.

MRC (Mekong River Commission). 2021. Basin Development Strategy 2021-2030 and MRC Strategic Plan 2021-2025. Vientiane: MRC Secretariat, https://doi: 10.52107/mrc.ajgonv.

Middleton, C. 2022. The political ecology of large hydropower dams in the Mekong Basin: a comprehensive review. *Water Alternatives* 15(2): 251-289, https://www.water-alternatives.org/index.php/alldoc/articles/vol15/v15issue2/668-a15-2-10/file.

Mother Nature Cambodia. (n.d.). Sand Mining Scam, https://mothernaturecambodia.org/sand-mining (accessed 7 January 2026).

Neuendorf, K.A. 2017. *The content analysis guidebook*. Thousand Oaks (CA): Sage Publications.

Park, E., Loc, H.H., Van Binh, D. and Kantoush, S. 2022. The worst 2020 saline water intrusion disaster of the past century in the Mekong Delta: impacts, causes, and management implications. *Ambio* 51(3): 691–699, https://doi.org/10.1007/s13280-021-01577-z.

Rogers, S., Fung, Z., Lamb, V., Gamble, R., Wilmsen, B., Wu, F. and Han, X. 2023. Beyond state politics in Asia's transboundary rivers: revisiting two decades of critical hydropolitics. *Geography Compass* 17(4): e12685, https://doi.org/10.1111/gec3.12685.

Rousseau, J.-F. and Habich-Sobiegalla, S. 2021. *The political economy of hydropower in Southwest China and beyond.* Cham: Palgrave Macmillan.

Rousseau, J.-F. and Marschke, M. 2023. (In)visible fluidities across sandscapes: Sand dredging and local socio-environmental impacts along the Red and Mekong Rivers. *The Extractive Industries and Society* 14(2023): 101261, https://doi.org/10.1016/j.exis.2023.101261.

Schmitt, R.J.P., Rubin, Z. and Kondolf, G.M. 2017. Losing ground – scenarios of land loss as consequence of shifting sediment budgets in the Mekong Delta. *Geomorphology* 294: 58–69, https://doi.org/10.1016/j.geomorph.2017.04.029.

Sopera, D. 2022. Corruption and water governance in the Mekong River Basin. *U4 Issue* 2022:12. Bergen: Chr Michelsen Institute: https://www.u4.no/publications/corruption-and-water-governance-in-the-mekong-river-basin.pdf (accessed 7 January 2026).

Soukhaphon, A., Baird, I.G. and Hogan, Z.S. 2021. The impacts of hydropower dams in the Mekong River Basin: a Review. *Water* 13(3): 265, https://doi.org/10.3390/w13030265.

Suykens, B., Marschke, M. and Rousseau, J.-F. 2023. Locating sand. *The Extractive Industries and Society* 15(2023): 101337, https://doi.org/10.1016/j.exis.2023.101337.

Tran, M. 2023. Quản lý khai thác, sử dụng, điều phối nguồn vật liệu san lấp theo hướng ưu tiên tiến độ các dự án. [*Manage exploitation, use, and coordination of landfill material sources in the direction of prioritizing project progress*]. Báo Tuổi Trẻ Thủ Đô, https://tuoitrethudo.vn/quan-ly-khai-thac-su-dung-dieu-phoi-nguon-vat-lieu-san-lap-theo-huong-uu-tien-tien-do-cac-du-an-233040.html (accessed 7 January 2026).

UNEP (United Nations Environment Program). 2019. Sand and sustainability: finding new solutions for environmental governance of global sand resources. Geneva: Grid-Geneva, UNEP Program, https://wedocs.unep.org/bitstream/handle/20.500.11822/28163/SandSust.pdf?sequence=1andisAllowed=y (accessed 7 January 2026).

UNEP (United Nations Environment Program). 2024. Global Resources Outlook 2024 – Bend the trend: pathways to a liveable planet as resource use spikes. Nairobi: UNEP, https://wedocs.unep.org/handle/20.500.11822/44901 (accessed 7 January 2026).

Valentine, P.C. 2019. Sediment classification and the characterization, identification, and mapping of geologic substrates for the glaciated Gulf of Maine seabed and other terrains, providing a physical framework for ecological research and seabed management. U.S. Geological Survey Scientific Investigations Report 2019–5073, https://doi.org/10.3133/sir20195073.

Vo, H. 2023. Hàng loạt công trình lớn vẫn lo thiếu cát: Cần phải có cơ chế đặc thù. [*A series of large projects are still worried about lack of sand: There needs to be a special mechanism*]. VietnamPlus, https://www.vietnamplus.vn/hang-loat-cong-trinh-lon-van-lo-thieu-cat-can-phai-co-co-che-dac-thu/898912.vnp (accessed 7 January 2026).

WWF (World Wildlife Fund). 2018. The sands are running out: how WWF sought to better understand the impact of dam construction and sand mining on sediment flows in the Mekong River basin, https://www.wwf.org.uk/sites/default/files/2018-04/180419_Mekong_sediment_CS-external.pdf (accessed 7 January 2026).

Xuan, T. 2023. Trữ lượng cát san lấp chỉ đáp ứng 70% nhu cầu xây đường cao tốc. [*Sand reserves only meet 70% of highway construction needs*]. Nhịp sống kinh tế Việt Nam and Thế giới, https://vneconomy.vn/tru-luong-cat-san-lap-chi-dap-ung-70-nhu-cau-xay-duong-cao-toc.htm (accessed 7 January 2026).

CHAPTER 8

The political ecologies of managing Cambodia's Tonle Sap fisheries

Mak Sithirith[1] and Kim Geheb[2]

Abstract

Situated in Cambodia, the Tonle Sap Lake (TSL) is the largest freshwater lake in Southeast Asia and the Mekong River Basin. It is a globally important biodiversity hotspot, while also displaying significant economic, livelihood, and food-production values. The lake and its resources are, however, under significant pressure from over-fishing and upstream development, particularly hydropower. In this chapter, we explore the management of the lake and its resources, arguing that the management frameworks in place reflect broader patterns of state interests and resource governance.

Keywords: Tonle Sap Lake, fisheries management, hydropower, Cambodia, state institutions.

Introduction

The Tonle Sap Lake (TSL) in Cambodia is a remarkable ecosystem. It plays a pivotal role in the country's and the regional social ecology through its immense fisheries and their considerable social and ecological value. The fishery is under significant stress, however, due to challenges in fisheries management, the influence of upstream factors such as hydropower development, and broader global factors, particularly climate change. The primary concern of this chapter is to respond to the question: why is it that the lake and the immense ecosystem services that it provides receive insufficient priority in state resource governance frameworks?

We address this question by drawing on concepts of state governance and resource control. We argue that the TSL has been managed in ways that reflect broader patterns of institutional organisation within the Cambodian state. While the lake and its resources have considerable socio-economic value to Cambodia, revenue from the lake is diffuse—comprising millions of daily, generally small, transactions. This diffuse character shapes how state institutions engage with the resource. Nevertheless, the state asserts itself upon the TSL's fisheries at strategic

[1] Cambodian Development Resource Institute, Phnom Penh, Cambodia.

[2] Landscape Alliance, Nairobi, Kenya.

moments—either to demonstrate authority, or to serve particular political objectives. In this chapter, we first provide a summary of how the Cambodian state's institutional systems function, drawing attention to how natural resources fit within these patterns. We then provide an introduction to the TSL, before considering its management over time. In its fourth section, we consider the ways in which the lake has been politicised—not necessarily because of its economic or ecological value to society, but in relation to broader political processes. The fifth section moves upstream, to consider hydropower development and its effects on the TSL. We focus on two dams to illustrate how fisheries considerations have not featured prominently in hydropower development decisions. The final section provides a discussion and concluding points.

This chapter is an essay based on combined author experience of 40+ years of working in and observing the Mekong, Cambodia, fisheries and the TSL. One of us (Sithirith) has extensive experience working on the TSL and in Cambodian fisheries management research.

The Cambodian state

The Cambodian state system is widely recognised as neo-patrimonial and centralised (Loughlin, 2024; Morgenbesser, 2019; Un, 2019). It is characterised by relationship networks that place the ruling party, the Cambodian People's Party (CPP), at the apex of power structures. The prime minister and his close allies serve as central figures in these networks, providing protection for loyal clients including private interest groups, relatives and kin, military forces, and tycoons (Baker and Milne, 2019). In return, these actors gain various benefits including favourable positions, contracts, or legal protections. In this way, the CPP has been able to establish networks that include those capable of providing the resources necessary to maintain political hegemony (Verver and Dahles, 2015).

Natural resources have featured prominently in these state institutional arrangements. Clients to these networks have been able to access logging, land, and mineral resources through preferential, informal arrangements. Clients are expected to provide financial contributions to entities associated with the ruling party. In recent years, these have included various party working groups and charitable organisations. Through these agencies, a wide variety of projects are financed, which include rural development projects. Critically, these are presented as gifts from specific leaders and figures, rather than as programmes of the formal state. This means that many functions conventionally performed by formal state institutions—such as financing of schools, temples, roads, and health centres—now fall into the domain of off-budget finance (Milne et al., 2015).

This system of financial networks extends through the administrative hierarchy. Government institutions (ministries, provincial or district administrations, and departments) remain necessary in order to provide formal legitimacy to governance. The budgets provided to these institutions, however, tend to be restricted. Budgets are often sufficient only to pay staff salaries, with limited resources for operational activities. Government officials are provided with various discretionary powers through law; it is through these that officials navigate their administrative responsibilities (Milne, 2015).

The limited operational budgets of government agencies mean that they are keen to expand their mandates, as this may bring additional resources or opportunities to exercise authority. Inter-ministerial competition in Cambodia can be intense, and such dynamics serve the central leadership through a divide-and-rule mechanism, which weakens the government as a whole whilst opening opportunities for particularly favoured ministers.

Cambodia has undertaken a variety of decentralisation reforms over recent decades. Authority remains highly centralised, however. Sub-national administrators—such as commune councils—have limited fiscal authority and no independent revenue-gathering capacity. In this way, Cambodia has managed to maintain centralised authority whilst developing the appearance of decentralisation (Ribot et al., 2006). Authority remains compressed into the highest levels of the state.

The Tonle Sap Lake

The TSL "defies uncomplicated characterization" (Roberts, 2015). In the dry season, the lake's aerial extent may drop to around 3,500 km^2 and a volume of about 1.5 km^3, whilst during the wet season, this increases to some 14,500 km^2 and a volume of 60–70 km^3 (Cowx et al., 2024; Kummu and Sarkkula, 2008) (Fig. 8.1). This significant variation in lake-level occurs because of the Mekong's 'flood pulse'. The lake connects to the Mekong mainstream via the Tonle River. When the monsoon commences, vast quantities of water surge through the system. As it enters the delta, the river floods and water backs up, before finding outlet via the Tonle River and into the TSL. Approximately 20 per cent of the Mekong's flood water is absorbed by the lake (Campbell et al., 2006; MRC, 2005). This serves to reduce flooding downstream. Once the pulse abates, the Tonle River reverses direction and drains the lake, ensuring that water remains available downstream and helping to prevent seawater intrusion into the Mekong Delta.

The flood pulse creates distinct ecological and hydrological benefits. Up and down the Mekong, and around the TSL, flooding makes available vast quantities of nutrients for fish. Unique flooded forests are inundated, opening up nutrient

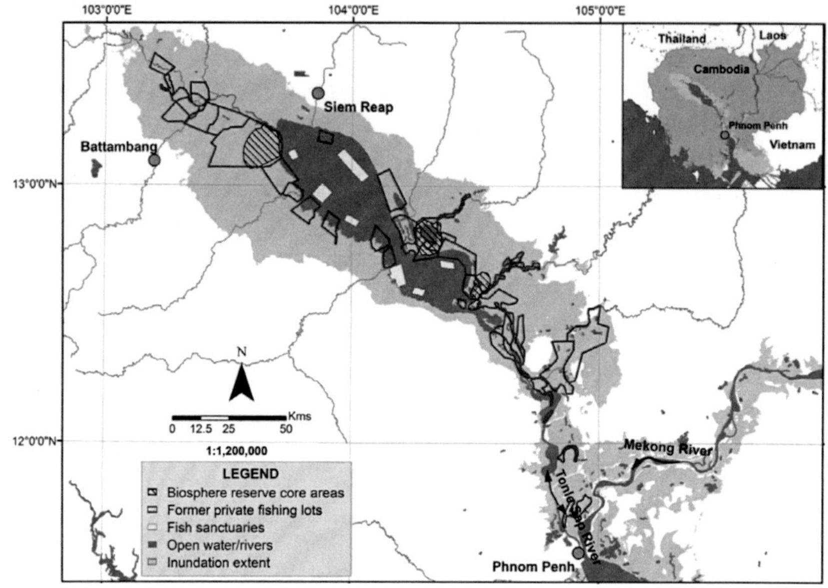

Figure 8.1: Map of the Tonle Sap, showing the flood extent and location of former fishing lots (Cooperman et al., 2012).

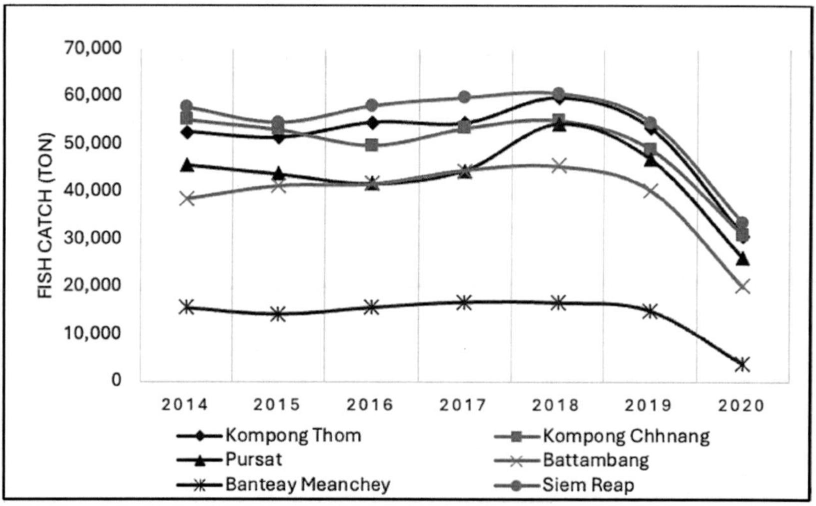

Figure 8.2: Fish catches from TSL provinces, 2014-2020 (FiA, 2021).

sources and creating important breeding places and refugia for fish. The flood pulse triggers one of the largest animal migrations on Earth (Hughes, 2024). From the TSL, fish migrate out into the Mekong system, and their young return. The flood pulse is the basis for one of the world's largest inland fisheries, immense biodiversity, and distinctive hydrological systems. The people on and around the lake have evolved complex relationships with these ecological cycles—synchronised to the rise and fall of the lake through recession agriculture on its floodplains, innovative fishing technologies, and floating villages (Sithirith and Grundy-Warr, 2025).

The TSL, writes Seiff (2022), is "a finely tuned ecological system. Pull a single thread and it will begin unravelling". There is considerable evidence that the TSL is experiencing significant stress. Fish catches have declined markedly. The *dai* (bag net) fishery, which is responsible for a large proportion of the TSL's fish landings, declined from 16,975 tonnes in 2018 to 9,900 tonnes in 2020 (Fig. 8.2). Populations of most of the system's fish species have declined by almost 90 per cent (Chevalier et al., 2023). Biodiversity throughout the Mekong system shows clear declines (Hughes, 2024; Sor et al., 2023).

The magnitude of the Mekong's flood pulse has decreased from an average of 49.7 billion m^3 (between 1962 and 1972) to 31.7 billion m^3 (between 2010 and 2018), a 36 per cent decline (Chua et al., 2022). The TSL's maximum flood water levels decreased by an average of 1.05 m between 2010 and 2019 as compared to 1996-2009 (Chua et al., 2022). The average maximum annual inundated area of the lake decreased by 16 per cent, from 11,917 km^2 (2001-2009) to 9,967 km^2 (2010-2020). Notably, 31 per cent of Tonle Sap's flooded forest has been lost since 1993 (Lohani et al., 2020; Chen et al., 2022). In 2016, approximately 250,000 hectares of flooded forest were lost to fires (Sassoon, 2016). Between 1995 and 2000, the TSL yielded some 0.65 $\pm$ 0.6 million tonnes of sediment annually downstream; since 2001, the lake has become a sediment sink, receiving 1.35 $\pm$ 0.7 million tonnes annually (Sok et al., 2021). In its floodplains, agricultural extension has encroached on grasslands and scrublands, such that their combined aerial extent declined by 27 per cent between 1993 and 2018 (Mahood et al., 2020).

The TSL has considerable economic value, generating an estimated USD 2 billion annually for the Cambodian economy (Manet, 2025)—approximately 8–10 per cent of Cambodia's GDP (ADB, 2022). Of Cambodia's approximately 16 million people (2019), six million people depend on fishing as their primary livelihood (NIS, 2008). On and around the TSL, some 1.7 million people live in 1,037 villages, of which 53 are floating, and 36 are located in its floodplain (Sithirith, 2011; 2015). Fish provides over 80 per cent of the protein in the national diet, and per capita annual consumption stands at 524 kg. Most harvested fish are migratory species (ADB, 2021).

Managing and zoning the Tonle Sap

Multiple government agencies are involved in the TSL's management. Its water is the purview of the Ministry of Water Resources and Meteorology (MOWRAM). The Tonle Sap Authority (TSA), which falls under MOWRAM, and the Cambodian National Mekong Committee (CNMC), which coordinates with the Mekong River Commission (MRC), are also involved. The Ministry of the Environment (MOE) chairs the Tonle Sap Biosphere Reserve (TSBR) Coordination Committee and is responsible for Cambodia's Ramsar sites. The Ministry of Agriculture, Forestry and Fisheries (MAFF) houses the Fisheries Administration and various agricultural extension agencies. The Ministry of Interior has relevance through the Royal Cambodian Police Force, as does the Ministry of Mines and Energy, given their oversight over hydropower development.

A system of fishing lots, applied to both the TSL and riverine fisheries, was established during the colonial period in 1908 (Cooke, 2011). Lots were auctioned to the highest bidder, who would then be allowed to fish these areas for a two-year period. Revenue derived from lot auctions, supplemented by taxes on fishing gear and customs duties on fish and related products, generated significant revenue. Whilst fisheries laws were formalised, their primary purpose was revenue generation rather than fisheries management (Degen and Thouk, 2000).

Following independence in 1953, little changed in the lot-based system. It was suspended under the Khmer Rouge but resurfaced in 1987 (Dina and Sato, 2014). By 2000, there were 56 fishing lots, covering about 507,731 hectares, in the TSL (Sithirith and Mathur, 2008; Yim and McKenney, 2004).

The lots yielded substantial catches with relatively limited fishing effort (De Lopez, 2002). Technically, subsistence fishers were permitted into lot areas during the closed fishing season, but the fences and armed guards ensured that such access was limited in practice. Lot concessions were expensive (ranging from USD 2,000 for a small sub-lease lot to USD 200,000 for a large sub-lease lot) and were allocated through informal processes (Chadwick et al., 2008; Degen et al., 2000; Sok, 2012). Conflict between lot concessionaires and small-scale fishers was widespread and sometimes resulted in fatalities (Degen et al., 2000).

In 1997, revenues from Tonle Sap lot auctions (USD 1.9 million) represented more than 65 per cent of the total fees collected from fisheries. The Fisheries Administration's annual budget, however, averaged less than USD 200,000 (De Lopez, 2002). In 2001, the fishing lot system was partially abandoned, a process finalised in 2012 when all fishing lots were annulled.

The post-lot approach yielded a plethora of new zone types (Table 8.1), frequently overlapping (Table 8.2). Several types of fishing zones were established, with the Fisheries Administration in charge of three main types:

- Fisheries Conservation Areas (FCAs): approximately 93,246 hectares (35 per cent) of abolished fishing lot areas were designated as 23 FCAs. Subsistence or small-scale fishing is allowed in FCAs.
- Fish sanctuaries: eight fish sanctuaries (24,680 hectares) were designated to provide refuges for fish broodstock. All fishing is banned from these areas.
- Community Fisheries (CFs): 222 CFs were established in the six provinces that share the lake, covering 538,739 hectares (Table 8.3). Only small-scale or subsistence fishing is permitted.
- Public fishing areas: waters not addressed by the above three zones are considered 'public', covering approximately 65 per cent of the lake area. These fisheries are characterised as open access, with no restrictions on who may fish or when.

Concerns about the loss of flooded forest were addressed via a separate zoning approach (TSA, 2011). Sub-decree no. 197 of 2012 designated three zones based on land use and vegetation characteristics, with different restrictions on human activity. The socio-economic activities in these zones are ostensibly monitored, with some activities being prohibited.

Zones also derive from the Tonle Sap's Biosphere Reserve (TSBR) status, obtained in 2001. This comprises core areas in which human activity is prohibited under Biosphere Reserve guidelines, surrounded by buffer and transition zones with different permitted activities. In negotiations between Cambodia and UNESCO, restrictions were somewhat relaxed, including in core areas. Additional zones are also associated with three Ramsar sites in the TSL – Prek Toal (registered 2015), Boueng Chhmar (registered 1999), and Stung Sen (registered 2018). These serve as important bird nesting and habitat areas and include significant tracts of flooded forest. The MOE is responsible for these areas.

With so many overlapping zones and variable mandates, the management framework has become complex. The diversity of zones, whilst intended to serve different conservation and management objectives, has resulted in an administrative framework in which coordination between agencies remains challenging.

Table 8.1: Cancelled fishing lots, public fishing, and fish conservation areas (DoF, 2001; FiA, 2012; and Sub-decrees of the fishing lots released for local peoples).

Province	Total fishing areas before 2000		Fishing lot areas cancelled in 2001		Fishing lot areas cancelled in 2012		Public fishing areas		Conservation area	
	# lots	Area (ha)	# lots	Area (ha)	# lots	Area (ha)	Area (ha)	%	Area (ha)	%
Banteay Meanchey	4	32,756	2	26,358	2	6,398	6,149	96.1	249	3.9
Battama-bang	12	146,532	3	43,814	9	102,718	50,169	48.8	52,550	51.2
Kampong Chhnang	19	62,256	7	17,171	12	45,085	35,125	77.9	9,959	22
Kampong Thom	7	127,126	0	57,773	7	69,353	49,166	73.7	17,503	26.2
Pursat	7	55,120	2	30,272	5	24,848	13,898	55.9	10,950	44.1
Siem Reap	7	83,941	4	61,216	3	22,725	20,690	91.0	2,035	8.9
Total	56	50,7731	18	236,604	38	271,127	175,197	65.3	93,246	34.7

Table 8.2: TSBR Core Areas, Ramsar Sites, and Fishery Conservation Area overlaps (MOE, 2022).

Zoning	TSBR Core Area (ha)	Ramsar Area (ha)	FCAs (ha)
Prek Toal	21,342	21,342	50,134
Boueng Chhmar	14,560	28,000	8,325
Stung Sen	6,365	9,293	921
Total	58,635	42,267	59,380

Table 8.3: TSL community fisheries by province (FiA, 2021).

Province	# Community fisheries	Membership	Area (ha)
Siem Reap	22	21,520	109,583
Kampong Thom	42	17,878	47,862
Kampong Chhnang	58	12,728	50,072
Pursat	35	8,901	88,350
Battambang	46	18,453	174,959
Banteay Meanchey	19	14,553	67,913
Total	**222**	**94,033**	**538,739**

The political dimensions of natural resource governance

The timing of significant changes to the lot system has coincided with important political moments. In 2000, the government announced that 8,000 hectares of lot area would be released to communities (Ratner, 2006). This release coincided with a period of intensive rural development efforts following electoral competition. The remaining fishing lots were released in 2012, an action that was linked to electoral competition and efforts to secure rural support.

In both instances, these decisions were sudden, and the decision-making process was restricted. Following the 2012 releases, observers noted that "powerful figures now privately *assume* that the Tonle Sap fishery is 'doomed' as a real source of commercial scale revenue" (Jones and Sok, 2015: 548, emphasis in original). The decision to abandon the fishing lots does not appear to have been intended to yield conventional fisheries management outcomes.

The release of the lots opened new opportunities for institutional manoeuvring. Multiple agencies began seeking control over different aspects of the lake, hence the creation of new zones of responsibility, primarily distributed between the Fisheries Administration, the MOE, and MOWRAM. With so many agencies with mandates for the lake, various coordinating institutions were proposed over the years. Initiatives to create overarching coordinating structures, however, faced difficulties from inter-ministerial dynamics, with agencies protective of their respective domains.

In 2006, the Law on Fisheries introducing Community Fishery (CF) zones was issued. CFs in Cambodia are generally understood to have been informed by broader global conversations around co-management (Sok et al., 2012; Sok and Yu, 2021). The Law on Fisheries, however, stipulated that all fisheries belong to the state. CFs do not hold powers of exclusion; they cannot license external fishers or seize illegal gear. These powers are retained by the Fisheries Administration. CFs are expected to raise their own funds to cover patrolling costs. In the absence of meaningful delegation of authority, the role and function of CFs remain ambiguous (Un, 2016).

Following the transition from lot-based to community fisheries, most observers characterise the fishery as open access (Chadwick et al., 2008; Ratner, 2006; Sithirith, 2015). Intensive fishing ensued as fishers sought to invest in larger equipment. This led to rapid overexploitation and shifted conflict from disputes with lot concessionaires to competition between fishers for access to diminishing fish stocks (Keskinen and Sithirith, 2009; Sour et al., 2013).

The 2006 Law on Fisheries provides formal legitimacy to fisheries management. It also established discretionary authorities that enable enforcement and compliance efforts. The zoning systems applied to the lake embody the territories within which authorities exercise these powers. Regular regulatory activities occur, often involving multiple agencies and law enforcement. Such activities reassert state authority over fisheries and natural resources. They also occur as sporadic crackdowns, however, rather than sustained management approaches.

The zone areas are not fixed. Periodic adjustments have been made to meet administrative or political objectives. Zone boundaries have occasionally been redrawn to allow specific communities to continue activities previously restricted. Such adjustments reflect the responsiveness of governance systems to local political dynamics, though this also means that the stability of management frameworks can be limited.

Upstream impacts: Hydropower and transboundary flows

The TSL is located at the receiving end of the Mekong's immense ecological and hydrological productivity. It is also at the receiving end of dominant regional economic and development narratives. The Mekong is extensively dammed, with thousands of irrigation and hydropower dams dotting its tributaries (Table 8.4).

The Mekong releases approximately 475 km^3 of water annually into the South China Sea. The basin as a whole has a theoretical hydropower potential of 70,872 megawatts (MW), 58 per cent of which lies in China (Geheb and Suhardiman, 2019). Cambodia lies mainly in the Mekong floodplains and lacks the steep valleys and gradients necessary for large-scale hydropower. Consequently, Cambodia's hydropower development has focused mainly on the country's southeast, in and around the Cardamom Mountains and tributary river systems.

Most of Cambodia's hydropower development in the Mekong Basin has focused on the Sesan and Srepok rivers, which, together with the Sekong River, form the '3S' system of transboundary tributaries. The Lower Sesan 2 (LSS2) dam (480 MW), commissioned in 2018, is located at the confluence of the Sesan and Srepok rivers, which are major fish migration tributaries. The dam is majority-owned by a major Chinese power producer, with one significant Cambodian shareholder.

The decision to proceed with the LSS2 was made in November 2012. The dam's environmental and social impact assessment anticipated significant changes to fisheries yields. A study by Ziv et al. (2012) anticipated that the LSS2 would cause a substantial reduction in the lower Mekong's fish biomass. In 2017, however, as the dam's construction neared completion, government officials indicated that there was no evidence that the LSS2 would impact Cambodia's fisheries.

Since the dam's commissioning in 2018, there have been no formal assessments of its individual impact on the 3S system's resources. Cumulatively, along with other 3S dams (mainly Vietnamese), however, the impacts are significant and affect sediments, biodiversity, fisheries, and water temperature (Meynell et al., 2024; Null et al., 2020; Souter et al., 2020).

Table 8.4: Hydropower dams in the Mekong River Basin (MERFI, 2025). Notes: installed capacity data excludes data from 11 Chinese and three Vietnamese commissioned dams, which are unavailable; installed capacity data for one under construction dam in China are excluded for the same reason.

	Commissioned		**Under construction**	
	# dams	Installed capacity (MW)	# dams	Installed capacity (MW)
Cambodia	3	494.1	1.0	80.0
China	45	23,539.1	5.0	126.0
Laos	79	10,095.2	13.0	2,567.7
Myanmar	1	66.0	0.0	0.0
Thailand	7	706.7	0.0	0.0
Vietnam	59	3,761.1	2.0	12.6
Totals	**194**	**38,662.1**	**21.0**	**2,786.3**

The LSS2's reservoir inundated 335 km^2 of land and resulted in the resettlement of almost 5,000 people. The process of public consultations, resettlement and compensation have been the subject of significant scrutiny and concern (HRW, 2021; Mahanty et al., 2023). Prior to inundation, the reservoir area had been heavily forested. The contract for reservoir clearance was awarded to a subsidiary of the main Cambodian shareholder. Logging activities were reported to have occurred outside the reservoir boundary, with logs allegedly being processed and transported through informal channels. Similar patterns of using hydropower projects to facilitate logging activities have been documented elsewhere in Cambodia (Käkönen and Thuon, 2019; Milne, 2015).

Upstream of Stung Treng, just across the border in Laos, is the Don Sahong hydropower dam (260 MW, later expanded to 325 MW). The dam is located in the Siphandon area, where the Mekong mainstream braids heavily. The plant sits on one channel, the Hou Sahong, which was previously well known as a major fish migration route.

As a mainstream dam, the Don Sahong was subject to the Mekong River Commission's (MRC) procedures for notification, prior consultation, and agreement (PNPCA). This process provides an opportunity to review and consider proposed dams, their designs, and associated impact assessments. The PNPCA process for the Don Sahong was complex, with Cambodia and Vietnam raising concerns about the project's environmental impacts and potential effects on downstream fisheries.

In January 2015, an impasse was reached in the PNPCA process. The MRC was forced to refer the matter to the MRC Council, comprising relevant ministers of member states. It is unclear if the Council ever formally convened, but the process was overtaken by events. Laos proceeded with construction. The dam was commissioned in 2020 as a 260 MW facility, with an additional 65 MW turbine installed and commissioned in 2024.

Cambodia has taken an active stance in critiquing proposed Lao mainstream dams, frequently raising concerns about potential impacts on Cambodian fisheries. This position frequently aligns with those of regional and international NGOs and the scientific community. Cambodia, however, has itself constructed the LSS2, which has delivered considerable impacts on social, hydrological, and ecological systems.

Hydropower development does create significant opportunities for engagement with international actors and funding, as well as associated commercial and administrative activities. The corporate arrangements surrounding hydropower development are often complex and involve numerous international partners. These arrangements create various interests and incentive structures that shape decision making.

The water levels of the TSL have become increasingly variable and difficult to predict. Research attributes these changes variously to climate change and localised anthropogenic activities such as channel incision, irrigation, sand-mining, and upstream hydropower plants (Hecht et al., 2019; Kallio and Kummu, 2021; Piman et al., 2016). Cumulative impact assessments in the Mekong Basin consistently show significant damage to fisheries and environmental systems (ICEM, 2010; IFC, 2021; MONRE, 2015; MRC, 2017).

The impacts of hydropower development on fisheries in the Mekong Basin are well documented in scientific literature. Hydropower development has, however, proceeded in Mekong countries with limited analysis of how fisheries losses are understood to be balanced against gains from electricity generation. None of the MRC's member states have provided detailed explanations of these trade-offs or clarified how benefits would be distributed across populations.

Discussion and conclusion

The Cambodian state operates through systems of institutional relationships and networks that have evolved over recent decades. To external observers, the formal Kingdom of Cambodia appears to operate according to traditional understandings of sovereignty and governance. These formal institutions have, however, been organised to serve particular interests and objectives within state institutions and networks. To understand governance outcomes—including the management of the TSL—it is necessary to understand how formal institutional frameworks interact with these broader patterns of state organisation.

The management of the TSL illustrates these dynamics. The lot-based system generated revenue but created tensions with small-scale fishing communities.

The release of the lots appears to have reflected political objectives of particular moments, rather than considered fisheries management strategy. Following the lot releases, multiple agencies sought to establish or expand mandates over different aspects of the lake, resulting in the complex zoning system that exists today.

While the Community Fisheries framework was presented as empowering fishing communities and supporting decentralisation, in practice it transferred few meaningful powers to local communities. The zoning systems, whilst designed to serve conservation objectives, operate in an environment where coordination between agencies remains challenging.

With respect to hydropower development, the record is instructive. Cambodia has advocated strongly for the consideration of fisheries impacts in transboundary dam negotiations, adopting positions aligned with regional and international scientific consensus. Simultaneously, Cambodia has proceeded with the construction of the LSS2, a project that impact assessments indicated would have substantial effects on fisheries and ecology. The apparent contradiction between Cambodia's stated concerns about upstream impacts and its own hydropower development suggests that decisions about hydropower are made on grounds other than fisheries considerations.

The arguments presented in this paper suggest that explaining fisheries management and hydropower development decisions requires understanding broader institutional and political dynamics. These include how state institutions are organised, the role of natural resources in institutional networks, and the various interests and incentive structures that shape governmental decisions. From this perspective, the TSL is undervalued and mismanaged not because of simple oversight or technical capacity gaps, but because of how the resource has been integrated—or rather, not integrated—into broader state governance and resource-allocation systems. Multiple agencies compete for mandates and resources; political leaders may use resource management decisions to serve broader political objectives; and hydropower development generates commercial interests that operate according to different logics than fisheries conservation. Understanding these dynamics is essential for any effort to improve the governance and sustainability of the TSL and its resources.

Acknowledgements

The authors gratefully acknowledge the many committed and conscientious government officials working throughout the Cambodian state system who are genuinely concerned about the sustainability of their country's natural resources and the wellbeing of its people. This analysis is not intended to diminish recognition of their efforts, but rather to identify structural dynamics that affect governance outcomes across institutional systems.

References

ADB (Asian Development Bank). 2021. ADB Provides Landmark Assistance to Cambodia's Marine Fisheries Sector. Phnom Penh: ADB Country Office, https://www.adb.org/news/adb-provides-landmark-assistance-cambodia-marine-fisheries-sector (accessed 6 January 2026).

ADB (Asian Development Bank). 2022. Cambodia agriculture, natural resources, and rural development sector assessment, strategy, and road map. Manila: Asian Development Bank, https://www.adb.org/sites/default/files/publication/718806/cambodia-agriculture-rural-development-road-map.pdf (accessed 6 January 2026).

Baker, J. and Milne, S. 2019. Cambodia's anti-corruption regime 2008–2018: a critical political economy approach. U4 Issue 2019:1. Bergen: U4 Anti-Corruption Research Centre, https://www.u4.no/publications/cambodias-anti-corruption-regime-2008-2018-a-critical-political-economy-approach.pdf (accessed 6 January 2026).

Baran, E. and Myschowoda, C. 2009. Dams and fisheries in the Mekong Basin. *Aquatic Ecosystem Health & Management* 12(3): 227–234, https://doi.org/10.1080/14634980903149902.

Campbell, I., Poole, C., Giesen, W. and Valbo-Jorgensen, J. 2006. Species diversity and ecology of Tonle Sap Great Lake, Cambodia. *Journal of Aquatic Sciences* 68(3): 355–373, https://doi.org/10.1007/s00027-006-0855-0.

Chadwick, M., Muanpong, J. and Sithirith, M. 2008. Sustaining Tonle Sap: an assessment of development challenges facing the Great Lake. Bangkok: Sustainable Mekong Research Network, https://www.academia.edu/3718625/Sustaining_the_Tonle_Sap_An_Assessment_of_Development_Challenges_Facing_the_Great_Lake (accessed 6 January 2026).

Chen, A., Chen, A., Varis, O. and Chen, D. 2022. Large net forest loss in Cambodia's Tonle Sap Lake protected areas. *Ambio* 51: 1889–1903, https://doi.org/10.1007/s13280-022-01704-4.

Chevalier, M., Ngor, P.B., Pin, K., Touch, B., Leh, S., Grenouillet, G. and Hogan, Z. 2023. Long-term data show alarming decline of majority of fish species in a Lower Mekong basin fishery. *The Science of the Total Environment* 891: 164624, https://doi.org/10.1016/j.scitotenv.2023.164624.

Cooke, N. 2011. Tonle Sap processed fish: from Khmer subsistence staple to colonial export commodity. In Tagliacozzo, E. and Chang, W.-C. (eds.), *Chinese circulations: capital, commodities, and networks in Southeast Asia*, 360–379. Durham (NC): Duke University Press.

Cooperman, M.S., So, N., Arias, M., Cochrane, T.A., Elliott, V., Hand, T., Lee, H., Holtgrieve, G.W., Kaufman, L., Koning, A.A., Koponen, J., Kum, V., McCann, K.S., McIntyre, P.B., Min, B., Ou, C., Rooney, N., Rose, K.A., Sabo, J.L. and Winemiller, K.O. 2012. A watershed moment for the Mekong: newly announced community use and conservation areas for the Tonle Sap Lake may boost sustainability of the world's largest inland fishery. *Cambodian Journal of Natural History* 2012 (2) 101-106, https://www.fauna-flora.org/publications/cambodian-journal-natural-history-december-2012/.

Cowx, I.G., Vu, A.V., Hogan, Z., Mallen-Cooper, M., Baumgartner, L.J., Quan Lai, T., Grill, G. and Sayer, C.A. 2024. Understanding the threats to fish migration: applying the Global Swimways Concept to the lower Mekong. *Reviews in Fisheries Science & Aquaculture* 33(2): 257–285, https://doi.org/10.1080/23308249.2024.2401018.

Degen, P. and Thouk, N. 2000. Historical, cultural and legal perspectives on the fishing lot system in Cambodia. In Ahmed, M. and Hirsh, P. (eds.), *Common property in the Mekong: issues of sustainability and subsistence*, 49–60. Penang and Sydney: WorldFish and the Australian Mekong Resources Center, University of Sydney.

De Lopez, T.T. 2002. Natural resource exploitation in Cambodia: an examination of use, appropriation, and exclusion. *The Journal of Environment & Development* 11(4): 355–379, https://doi.org/10.1177/1070496502238662.

Dina, T. and Sato, J. 2014. Is greater fishery access better for the poor? Explaining de-territorialisation of the Tonle Sap, Cambodia. *The Journal of Development Studies* 50(7): 962–976, https://doi.org/10.1080/00220388.2014.909027.

DoF (Department of Fisheries). 2001. The data of the release of fishing lot areas in Cambodia. Phnom Penh: DoF, Ministry of Agriculture, Forestry and Fishery.

FiA (Fishery Administration). 2021. Fishery database. Phnom Penh: FiA, Ministry of Agriculture, Forestry and Fishery.

Geheb, K. and Suhardiman, D. 2019. The political ecology of hydropower in the Mekong River Basin. *Current Opinion in Environmental Sustainability* 37: 8–13, https://doi.org/10.1016/j.cosust.2019.02.001.

Hecht, J.S., Lacombe, G., Arias, M.E., Dang, T.D. and Piman, T. 2019. Hydropower dams of the Mekong River basin: a review of their hydrological impacts. *Journal of Hydrology* 568: 285–300, https://doi.org/10.1016/j.jhydrol.2018.10.045.

Hughes, K. 2024. The Mekong's forgotten fishes and the emergency recovery plan to save them. Gland: World Wildlife Fund, https://wwfasia.awsassets.panda.org/downloads/final-mekong-forgotten-fishes-report--web-version-.pdf (accessed 6 January 2026).

HRW (Human Rights Watch). 2021. Underwater: human rights impacts of a China Belt and Road project in Cambodia. New York: HRW, https://www.hrw.org/sites/default/files/media_2021/08/cambodia0821_web.pdf (accessed 6 January 2026).

Jones, R.W. and Sok, S. 2015. Impacts and implications of deep fisheries reforms on the governability of small-scale fisheries in Tonle Sap Lake, Cambodia. In Jentoft, S. and Chuenpagdee, R. (eds.), *Interactive governance for small-scale fisheries*, 539–557. MARE Publications Series 13. Cham: Springer.

Käkönen, M. and Thuon, T. 2019. Overlapping zones of exclusion: carbon markets, corporate hydropower enclaves and timber extraction in Cambodia. *The Journal of Peasant Studies* 46(6): 1192–1218, https://doi.org/10.1080/03066150.2018.1474875.

Keskinen, M. and Sithirith, M. 2009. Tonle Sap Lake and its management: the diversity of perspectives & institutions. Mekong Water Allocation Project (PN67) Study, Revised version 31 July 2009, https://www.academia.edu/3718631/Tonle_Sap_Lake_and_its_management_the_diversity_of_perspectives_and_institutions (accessed 6 January 2026).

Kummu, M. and Sarkkula, J. 2008. Impact of the Mekong River flow alteration on the Tonle Sap flood pulse. *Ambio* 37(3): 185–192, https://doi.org/10.1579/0044-7447(2008)37[185:iotmrf]2.0.co;2.

Loughlin, N. 2024. *The politics of coercion: state and regime making in Cambodia*. Ithaca (NY): Cornell University Press.

Mahanty, S., Chann, S. and Suong, S. 2023. The emotional life of rupture at Cambodia's Lower Sesan 2 hydropower dam. *Environment and Planning E: Nature and Space* 7(1): 330–352, https://doi.org/10.1177/25148486231162087.

Mahood, S.P., Poole, C.M., Watson, J.E.M., MacKenzie, R.A., Sharma, S. and Garnett, S.T. 2020. Agricultural intensification is causing rapid habitat change in the Tonle Sap Floodplain, Cambodia. *Wetlands Ecology and Management* 28(2020): 713–726, https://doi.org/10.1007/s11273-020-09740-1.

Manet, S. 2025. Tonle Sap Lake contributes $2 billion annually to economy. *Khmer Times*, 2 January 2025, https://www.khmertimeskh.com/501617685/tonle-sap-lake-contributes-2-billion-annually-to-economy/ (accessed 6 January 2026).

MERFI (Mekong Region Futures Institute). 2025. Dataset on the dams of the Greater Mekong. Bangkok, Mekong Region Futures Institute: https://www.merfi.org/mekong-region-dams-database (accessed 6 January 2026).

Meynell, P.-J., Metzger, M.J. and Stuart, N. 2024. Assessing the impacts of changing connectivity of hydropower dams on the distribution of fish species in the 3S rivers, a tributary of the lower Mekong. *Water* 16(11): 1505, https://doi.org/10.3390/w16111505.

Milne, S. 2015. Cambodia's unofficial regime of extraction: illicit logging in the shadow of transnational governance and investment. *Critical Asian Studies* 47(2): 200–228, https://doi.org/10.1080/14672715.2015.1041275.

Milne, S., Kimchoeun, P. and Sullivan, M. 2015. Shackled to nature? The post-conflict state and its symbiotic relationship with natural resources. In Milne, S. and Mahanty, S. (eds.), *Conservation and development in Cambodia: exploring frontiers of change in nature, state and society*, 28–50. London: Routledge-Earthscan.

Morgenbesser, L. 2019. Cambodia's transition to hegemonic authoritarianism. *Journal of Democracy* 30(1): 158–171, https://doi.org/10.1353/jod.2019.0012.

Null, S.E., Farshid, A., Goodrum, G., Gray, C.A., Lohani, S., Morrisett, C.N., Prudencio, L. and Sor, R. 2020. A meta-analysis of environmental tradeoffs of hydropower dams in the Sekong, Sesan, and Srepok (3S) rivers of the lower Mekong Basin. *Water* 13(1): 63, https://doi.org/10.3390/w13010063.

Piman, T., Cochrane, T.A. and Arias, M.E. 2016. Effect of proposed large dams on water flows and hydropower production in the Sekong, Sesan and Srepok rivers of the Mekong Basin. *River Research and Applications* 32(10): 2095–2108, https://doi.org/10.1002/rra.3045.

Ratner, B. 2006. Community management by decree? Lessons from Cambodia's fisheries reform. *Society and Natural Resources* 19(1): 79–86, https://doi.org/10.1080/08941920500323344.

Ribot, J.C., Agrawal, A. and Larson, A.M. 2006. Recentralizing while decentralizing: how national governments reappropriate forest resources. *World Development* 34(11): 1864–1885, https://doi.org/10.1016/j.worlddev.2005.11.020.

Roberts, A.S. 2015. Lost in transition: landscape, ecological gradients and legibility on the Tonle Sap floodplain. In Milne, S. and Mahanty, S. (eds.), *Conservation and development in Cambodia: exploring frontiers of change in nature, state and society*, 53–74. London: Routledge-Earthscan.

Sassoon, A.M. 2016. Tonle Sap forests razed by fire. *The Phnom Penh Post*, 20 May 2016, https://www.phnompenhpost.com/national/tonle-sap-forests-razed-fire (accessed 6 January 2026).

Seiff, A. 2022. *Troubling the water: a dying lake and a vanishing world in Cambodia*. Lincoln (NE): Potomac Books.

Sithirith, M. 2011. Political geographies of the Tonle Sap: power, space and resources. PhD Thesis. Singapore: National University of Singapore, https://www.academia.edu/3718644/Political_Geographies_of_the_Tonle_Sap_Power_Space_and_Resources (accessed 6 January 2026).

Sithirith, M. 2015. Tonle Sap Lake, Cambodia: a perspective on its transborder hydrological and resources governance. *Lakes and Reservoirs: Research and Management* 20: 187–205, https://doi.org/10.1111/lre.12105.

Sithirith, M. and Grundy-Warr, C. 2025. The social flood pulse and socio-ecological transformation of the Tonle Sap. *Singapore Journal of Tropical Geography* 46(1): 67–94, https://doi.org/10.1111/sjtg.12573.

Sithirith, M. and Mathur, V. 2008. Entitlements and the community fishery in the Tonle Sap: is the fishing lot system still an option for inland fisheries management? In Chadwick, M., Juntopas, M. and Sithirith, M. (eds.), Sustaining Tonle Sap: an assessment of development challenges facing the Great Lake, 99–121. Bangkok: Stockholm Environment Institute.

Sok, S. 2012. State building in Cambodia. PhD Thesis. Melbourne: Deakin University, https://dro.deakin.edu.au/ndownloader/files/37443577/1 (accessed 6 January 2026).

Sok, S., Wang, F. and Chhinh, N. 2022. Political participation and small-scale fishery management in the Tonlé Sap, Cambodia. *International Journal of Water Resources Development* 38(4): 616–633, https://doi.org/10.1080/07900627.2021.1987201.

Sok, S. and Yu, X. 2021. Co-management of small-scale fishery in the Tonle Sap Lake, Cambodia. *Regional Sustainability* 2(1): 1–11, https://doi.org/10.1016/j.regsus.2020.12.002.

Sok, S., Yu, X. and Wong, K.W. 2012. Impediments to community fisheries management: some findings in Kompong Pou commune, Krakor District in Cambodia's Tonle Sap. *Singapore Journal of Tropical Geography* 33(3): 398–413, https://doi.org/10.1111/sjtg.12007.

Sor, R., Ngor, P.B., Lek, S., Chann, K., Khoeun, R., Chandra, S., Hogan, Z.S. and Null, S.E. 2023. Fish biodiversity declines with dam development in the lower Mekong Basin. *Nature: Scientific Reports* 13: 8571, https://doi.org/10.1038/s41598-023-35665-9.

Souter, N., Shaad, K., Vollmer, D., Regan, H.M., Farrell, T.A., Arnaiz, M., Meynell, P.-J., Cochrane, T.A., Arias, M.E., Piman, T. and Adelman, S.J. 2020. Using the Freshwater Health Index to assess hydropower development scenarios in the Sesan, Srepok and Sekong River Basin. *Water* 12(3): 788, https://doi.org/10.3390/w12030788.

Sour, K., Kosal, M. and Sovannarith, S. 2013. Fishery reforms on the Tonle Sap Lake: risks and opportunities for innovation. *Cambodia Development Review* 17(2): 1–4, https://www.cdri.org.kh/storage/pdf/cdr13-2e-1_1617683513.pdf.

Un, B. 2016. 'Fishing-farming' communities around Tonle Sap Lake, Cambodia: fisheries reforms, natural resource management and changing livelihood patterns. In Blake, D.J.H. and Robins, L. (eds.), *Water governance dynamics in the Mekong Region*, 211–238. Petaling Jaya and Chiang Mai: Strategic Information and Research Centre and the Unit for Social and Environmental Research.

Un, K. 2019. *Cambodia: return to authoritarianism*. Cambridge: Cambridge University Press.

Verver, M. and Dahles, H. 2015. The institutionalisation of *Oknha*: Cambodian entrepreneurship at the interface of business and politics. *Journal of Contemporary Asia* 45(1): 48–70, https://doi.org/10.1080/00472336.2014.891147.

Yim, C. and McKenney, B. 2004. Domestic fish trade: a case study of fish marketing from the Great Lake to Phnom Penh. CDRI Working Paper 29. Phnom Penh: Cambodia Development Resource Institute, https://cdri.org.kh/storage/pdf/wp29e_1617794682.pdf (accessed 6 January 2026).

Ziv, G., Baran, E., Rodríguez-Iturbe, I. and Levin, S.A. 2012. Trading-off fish biodiversity, food security, and hydropower in the Mekong River Basin. *Proceedings of the National Academy of Sciences* 109(15): 5609–5614, www.pnas.org/cgi/doi/10.1073/pnas.1201423109.

CHAPTER 9

Voicing the Mekong: Listening to humans, rivers, and spirits in the politics of hydropower

Kanokwan Manorom[1] and Kim Geheb[2]

Abstract

This chapter examines how different 'voices' are heard—and silenced—in Mekong hydropower governance. Drawing on Latour's 'parliament of things' and voice studies scholarship, we argue that consultation mechanisms privilege techno-scientific and state speech while excluding community knowledge, the river's material presence, and cosmological registers through which villagers understand the Mekong. Based on ethnographic research with communities facing the Ban Kum/Salawan Dam and those affected by Nam Theun 2, we present villagers' articulations of multiple voices—affective, sonic, corporeal, ecological, and spiritual. Situating these within legal frameworks recognising rivers as rights-bearing entities, we propose 'deep listening' as a governance methodology that accommodates multiple ontologies of voice and creates space for the river and its communities as genuine stakeholders.

Keywords: Mekong River, hydropower, voice, rights of nature, deep listening, animism

Introduction: The problem of voice in Mekong governance

A key concern for Bruno Latour was the separation of humans from nonhumans—the former relegated to the domain of politics, the latter to science (Latour, 1993). This 'Great Divide', as Latour termed it, has profound consequences for how we govern shared environments. In his later work, Latour (2004) proposed a 'parliament of things' in which nonhuman entities would be granted political voice, represented alongside humans in collective decision making about matters of common concern. The Mekong River, shared by six nations and sustaining some 60 million people, presents a compelling case for such expanded political representation. Yet the governance of Mekong hydropower development reveals

[1] Ubon Ratchathani University, Thailand.

[2] Landscape Alliance, Nairobi, Kenya.

precisely the exclusions that Latour's work diagnoses: certain voices are amplified while others are systematically silenced.

This chapter examines how different 'voices' are heard—and not heard—in the politics of dam construction on the Mekong River. We take voice seriously in multiple registers: as a literal acoustic phenomenon, as political speech, as metaphorical representation, as affective expression, and as cosmological presence. Drawing on ethnographic research with communities facing the proposed Ban Kum/Salawan Dam, and earlier fieldwork with communities affected by Nam Theun 2, we present the voices of villagers, the river, and the spiritual beings that inhabit the Mekong world. These voices, we argue, deserve recognition not merely as 'stakeholder input' to be processed through existing consultation mechanisms, but as articulations of fundamentally different ways of knowing and being with rivers that challenge the ontological assumptions underlying contemporary hydropower governance.

The loudest and most powerful voices in Mekong hydropower governance are those of states, which set energy policy; electricity utilities, which purchase power; and private firms, which possess the capital, technical knowledge, and political connections to construct dams. Power Purchase Agreements are signed before consultations conclude; the Mekong River Commission's (MRC) Procedures for Notification, Prior Consultation and Agreement (PNPCA) proceed even as communities voice opposition. International investors and electricity utilities function as 'ghost stakeholders'—establishing financial hierarchies through PPAs that render local voicing moot before it even begins, since legal and financial obligations to capital outweigh ecological and moral obligations to riverine communities (Dore *et al.*, 2012). In Thailand, electricity reserves already exceed demand by 40 per cent, yet new agreements continue to be signed for power from Lao dams (Permphongsacharoen, 2018). The voices of villagers, civil society, academia, and the media—while present—have had little influence on shaping the Mekong's developmental trajectory. The voices of the river itself, and of the spirits that villagers understand as inhabiting it, receive no hearing at all.

We structure this chapter as follows. First, we develop a theoretical framework that moves from 'voice' as noun to 'voicing' as verb, drawing on Latour's political ecology and recent work in voice studies. Second, we present ethnographic material organised around multiple registers of voice: affective, sonic, corporeal, ecological, cosmological, and gendered. Third, we analyse the mechanisms through which these voices are excluded from governance processes. Finally, we propose 'deep listening' as a methodology for more just and ecologically attuned Mekong governance, drawing on emerging legal frameworks that recognise rivers as entities with rights.

The ethnographic material presented here draws on fieldwork conducted between February and August 2021 (15 interviews), and May–November 2025 (13 interviews). Research with the communities facing the Ban Kum/Salawan Dam involved interviews and focus group discussions with villagers in both Thailand and Laos. The earlier research with Bru women affected by Nam Theun 2 was conducted in 2024. Interviews were conducted in Thai and Lao, with translation assistance where necessary. Given the sensitivity of discussing spiritual beliefs and opposition to state-backed projects, we have anonymised all participants and obtained informed consent for the use of their accounts.

Theoretical framework: From voice to voicing

Latour's parliament of things

In Politics of Nature, Latour (2004) argues that modern politics has been crippled by its exclusion of nonhumans from the collective. Science claims to speak for nature—to reveal its objective truths—while politics is relegated to the messy business of managing human interests. This division, Latour argues, is both epistemologically false and politically disabling. Scientists do not simply discover facts about nature; they actively construct representations of nonhuman entities through instruments, experiments, and interpretive frameworks. Meanwhile, the exclusion of nonhumans from political deliberation means that the entities most affected by collective decisions—rivers, forests, species, ecosystems—have no voice in those decisions.

Latour's solution is to compose a 'collective' in which humans and nonhumans participate together. This requires abandoning the fiction that science provides unmediated access to nature's voice. Instead, we must recognise that nonhumans always speak through intermediaries—scientists, certainly, but also farmers, fishers, indigenous peoples, and others whose practices bring them into intimate relationship with nonhuman entities. The question becomes: whose representations of nonhumans will be admitted to political deliberation, and on what terms?

Applied to Mekong governance, Latour's framework illuminates the politics of representation at stake in hydropower decisions. The Mekong River does not speak for itself. It speaks through those who claim to represent it. Currently, dominant representations come from hydrologists, engineers, and environmental consultants who model the river's flows, sediment transport, and fish migrations in terms amenable to cost–benefit analysis. But villagers along the Mekong offer different representations: the river as kin, as body, as person, as a home of spirits. These representations are not merely 'cultural beliefs' to be noted and set aside; they are knowledge claims about what the river is and what it needs, grounded in generations of intimate interaction.

Voice studies and the politics of voicing

Recent scholarship in voice studies offers resources for thinking about how voice operates across sonic and political registers. Eidsheim and Meizel (2019) argue that voice is simultaneously material and metaphorical: a physical phenomenon produced by bodies and transmitted through air, and a figure for agency, identity, and political presence. This duality is captured in our everyday language: we speak of 'having a voice' in decisions, of 'giving voice' to the voiceless, of 'raising our voices' in protest. Voice, in these usages, names the capacity to be heard and to matter in collective deliberation.

Revill (2021) extends this analysis to environmental politics, arguing for a conception of 'voicing' as an emergent property of socio-material assemblages rather than a capacity possessed by individual subjects. Drawing on Latour's later work, Revill critiques the assumption that "human language and the rational speaking subject" should serve as the benchmark for political participation. If voice is understood as requiring articulate speech in recognisable language, then rivers, ecosystems, and nonhuman species are necessarily excluded. But if voicing is understood as the capacity to make a difference—to affect and be affected by others in consequential ways—then the Mekong River voices its presence constantly: through its flows and floods, its fish and sediments, its sounds and silences.

Kanngieser (2012) develops what she calls a 'sonic geography of voice' that attends to the material properties of vocal sound—its pitch, timbre, rhythm, and affective resonance. This approach proves valuable for analysing how villagers along the Mekong describe the river's sounds, and how those sounds have changed with dam construction upstream. The river's voice, in this literal sense, has been altered: its rhythms disrupted, its seasonal patterns made unpredictable. Attending to this sonic dimension reveals something that purely political or metaphorical analyses of voice might miss: the embodied, sensory relationship between villagers and river that is at stake in hydropower development.

Spivak revisited: What counts as speech?

Gayatri Spivak's famous question—'Can the subaltern speak?'—is often misread as asking whether marginalised people are capable of speech (Spivak, 1988). Her actual argument is more radical: the subaltern cannot speak within hegemonic discourse structures because those very structures determine what counts as speech, knowledge, and legitimate claims. The problem is not that subaltern voices are too quiet; the problem is that the criteria for audibility are set by dominant groups. Reading Spivak alongside Latour complicates any project of ontological inclusion: where Latour seeks to expand representation to nonhumans, Spivak reminds us that representation itself is already shaped by power, and that inclusion without transformation of epistemic criteria risks reproducing the very exclusions that it aims to address.

This stronger version of the argument serves our purposes well. When villagers describe the Mekong as a mother, when they speak of its happiness or illness, when they invoke the nagas who created the river, they are not speaking a language that hydropower governance is designed to hear. The PNPCA process assumes a particular kind of speaking subject: one who articulates interests in terms of impacts and compensation, who engages with technical assessments and environmental management plans. Villagers can learn to speak this language—and many have, with support from civil society organisations—but doing so requires translating their knowledge into terms that inevitably distort it. The question of the Mekong's personhood, its spiritual inhabitants, its right to flow freely—these become, at best, 'cultural concerns' to be 'taken into account' rather than ontological claims that challenge the very framework of governance.

The voices of the Mekong: Ethnographic encounters

The ethnographic material presented here derives from research with communities facing the proposed Ban Kum/Salawan Dam—a planned large-scale dam to be located on the Mekong mainstream between Thailand and Laos, in the areas of Ban Kum and Ban Ta Mui in Ubon Ratchathani Province, Thailand, and villages in Champasak Province, Laos. The MRC estimates that the dam will produce 1,872 MW of electricity, displacing some 2,570 people across seven Thai communities and six Lao villages (MRC, 2010). The project was controversial a decade ago when widespread protests led to its suspension, but it has since been reconsidered. Companies involved include Energy Absolute PCL, which signed a Memorandum of Understanding to assess environmental impacts; and Italian-Thai Development, the proposed construction contractor. We also draw on earlier research with Bru women affected by Nam Theun 2 in Laos (Manorom, 2017).

We organise villagers' articulations around six registers of voice, while recognising that these registers are deeply interconnected in practice.

Affective voices: Intimacy, happiness, and belonging

> *I have had bonds with the rivers and the streams for years and it makes me feel happy.*

> *Originally, we could drink water from the Mekong River and play in the water with peace of mind.*

> *I was born with the belief that the Mekong River is ours. Living in our village makes us the happiest people in our country. When we go into the water, there are fish, shellfish, and shrimp for us to catch.*

> *The fish, the gardens and the Mekong River are ours. They don't belong to someone else. They are our property.*

Villagers along the Mekong speak of their relationship with the river in terms of intimacy, happiness, and belonging—affective registers that are difficult to translate into the language of Environmental Impact Assessment (EIA). These are not merely sentimental attachments; they name a mode of being-in-the-world that is fundamentally relational.

These statements articulate a sense of ownership that is not possessive individualism but relational belonging. The river is 'ours' not because villagers hold title to it, but because they have grown up with it, depend upon it, and identify with it. This sense of belonging grounds 'entitlement'—the legitimate claim to access and benefit from a resource (Sen, 1981). When villagers say that the Mekong is theirs, they are claiming a history of association.

The prospect of dam construction transforms these affective registers from happiness to anxiety:

> *Our happy laughs and smiles turned into unhappy ones because once the dam is built, we will have to leave the village. I'm worried. Not a lot of people know how to deal with it and I don't know how to fight for it.*

> *If we have to evacuate from the village, we, the villagers will suffer.*

Sonic voices: The river's sounds

Villagers attend to the river's literal voice—its sounds—in ways that reveal an intimate acoustic relationship.

> *The Mekong is happy. When we see the Mekong, when we look at it, when we touch it, we see the Mekong flowing, we are happy. The Mekong is free, the Mekong can do anything. The Mekong is relaxing and calming. The pattern that the wind blows, the coolness of the water and the wind are natural, not man-made.*

> *When the Mekong River recedes, the Mekong flows continuously and the sound is pleasant to listen to. It's a sound that makes you want to sleep. It's soft and loud, but it's not a weak sound. It sounds like the voice of a strong person. As for when the Mekong River rises, her voice is very strong and loud from the water that hits the rocks.*

This sonic dimension connects directly to Revill's (2016) work on sonic geographies and the political agency of sound.

These descriptions reveal sophisticated attention to the river's acoustic properties: its rhythm, volume, and affective qualities. The river has seasonal voices—softer during the dry season recession, stronger during the rise. These voices are gendered (the river is 'she' with a 'strong' voice) and characterised in terms of personality ('the voice of a strong person'). The river's voice is not metaphorical here; villagers are describing what they literally hear, and what that hearing means to them.

Dams upstream have already begun to alter this sonic landscape. The eleven dams in China's Yunnan Province, completed before 2023, have disrupted the seasonal rhythms that gave the river's voice its predictable patterns. Villagers can no longer anticipate flows, tides, or fish migrations. The river still speaks, but its voice has become erratic, unpredictable—no longer the reliable companion whose rhythms structured village life.

Corporeal voices: The river as body

Villagers consistently describe the Mekong using bodily metaphors—the river as a living organism, with tributaries as limbs and the main channel as its trunk.

> *The small tributaries are arms of the Mekong River. The Mekong River is the body. When the tide rises, the Mekong River flows into its tributaries. I feel that bathing in the tributaries is like bathing in the Mekong. The tributaries are a part of the Mekong River and they are connected.*

This corporeal imagination is not a mere figure of speech; it articulates an understanding of the river as an integrated whole, vulnerable to injury.

The bodily understanding has direct implications for how villagers conceptualise dam impacts:

> *If it is built [Ban Kum Dam], the life [of the Mekong River] will no longer be private because it has been invaded. Previously, it was like we and the Mekong were siblings and friends. We rely on our friend. We depend on each other. We catch fish and other animals from the river. When Ban Kum Dam is built, suffering will come into our life. It'll probably make you sick. It's like blood that doesn't circulate in your body and causes you disease and sickness. Those who depend on the Mekong will also suffer, including people, shellfish, crabs, and fish. If we are ill, we have no strength and we have no power.*

The dam, in this account, is an invasion of the river's body that disrupts its circulatory system, causing illness that spreads to all who depend upon it. The metaphor of blocked blood circulation captures something that technical discourse on sediment transport also recognises—dams trap the nutrients that sustain downstream ecosystems—but frames it in terms of suffering and sickness rather than reduced sediment loads.

Villagers extend this bodily imagination to diagnose the river's current condition:

> *The river is sick. The Mekong is dying.*

> *The Mekong River is now seriously ill. If it's clogged by dams, the Mekong River won't exist. There will only be the river full of dams.*

> *The Mekong River will disappear, it will be called by the name of each dam here and there instead. The Mekong River will disappear because it is dead.*

> *Today, they build dams on different parts of the Mekong River, which means they are crushing the Mekong River. There is no abundance in the river. The ecosystem is in bad condition, unlike in my parents' times when the Mekong River was very abundant.*

The fear is not merely that dams will reduce fish catches or alter water levels, but that the Mekong as a coherent entity—as a living body—will cease to exist, replaced by a series of reservoirs bearing the names of the dams that created them.

Ecological voices: Sediment, fish, and the river's material speech

The Mekong's nonhuman constituents—water, sediment, sand, gravel, stones, rapids, rock pits, soil, nutrients, forests, plants, fish, shrimp, shellfish, frogs—are, in Latourian terms, actants whose presence and behaviour make differences in the world. Villagers speak for these entities, giving voice to the 'silent people' of the Mekong ecosystem.

The tributaries in the area where the Ban Kum/Salawan Dam will be built are crucial to this ecological assemblage. Huai Ta Mui, Huai Na Saen, and Hong Bon are not merely water channels but spawning grounds for fish during the high tide season, sources of livelihood for villagers, and sites for vegetable cultivation when the water recedes. Women, in particular, rely on these small streams for hunting aquatic animals—frogs, fish, shrimp, and shellfish—because they are easily accessible. The dam would submerge these tributaries, eliminating their ecological functions and the livelihoods that they support.

Scientific research confirms what villagers observe: dams trap sediment that would otherwise fertilise downstream floodplains, provide nutrients for fish, and replenish riverbanks (Baran et al., 2015; Schmitt et al., 2019). The eleven dams in China's Yunnan Province have already reduced sediment transport, with cascading effects on fish populations and agricultural productivity (e.g., Soukhaphon et al., 2021). Villagers articulate these impacts in their own terms:

> *The future of the river will not be the same if the dam is constructed. The water will be contaminated.*

> *How do we manage the river? I think I don't want a dam to be built anymore because with the existence of the dam, the water won't recede and it will be difficult for me to make a living, and any agricultural activities along the Mekong River won't be possible. Crabs and fish are also becoming harder to find.*

The uncertainty that dams introduce is profound. Villagers can no longer predict water levels, tides, fish migration, or natural food availability for fish. Sediment transport—the nutrients that sustain crops along the riverbanks—has become unpredictable. Even water traditions such as Songkran are disrupted when the river's behaviour no longer follows seasonal patterns (Manorom, 2022; Johnson, 2020). This is what Haraway (1997) calls 'situated knowledge' made unreliable: the practical expertise accumulated over generations becomes useless when the river no longer behaves as it always has.

Cosmological voices: Nagas, spirits, and the sacred river

The voices discussed thus far—affective, sonic, corporeal, ecological—are, in principle, accessible to outside observers. But villagers also speak of voices that exist in the realm of thoughts, worldviews, legends, chronicles, and beliefs: the spirits, ancestral beings, forest gods, mountain spirits, or nagas that populate the Mekong world. This dimension, which might be called animism, functions as a form of social organisation maintaining relationships between local people and supernatural beings (Yvonne, 2017). These nonhuman beings are part of villagers' lives; their voices are crucial elements of social life, kinship, mental health, and community belonging.

The Mekong itself, in local cosmology, was created by nagas—serpent spirits whose actions shaped the physical landscape. Wongthet (2023) recounts the legend:

> The Mekong River was created because of the nagas fighting against each other… [T]he nagas living in Nong Sae [in Yunnan Province, southern China] quarrelled with one another. A group of Nagas had to escape so they dug up the land to pass through and the channels became the Mekong River.

The legend recounts how Phaya Srisatta Nakharat and Phraya Sutta Naga, friends who lived in Nong Krasae Luang, fell into conflict over a gift of porcupine meat. After seven days and seven nights of battle, the defeated party fled to the northeast, and the channels which they dug became the Mekong. The river, in this account, is the product of nonhuman agency—the material trace of a quarrel between spirit beings. This origin story continues to shape how villagers understand their relationship with the river and their obligations to its spiritual inhabitants.

This cosmological dimension reflects what Herzfeld (2004) calls 'cultural intimacy'—shared systems of meaning that outsiders may not understand but that bind communities together. People outside the culture may not grasp the significance of these beliefs because local people build relationships with one another through the creation of a shared value system of cultural objects. Yet these beliefs are not peripheral to the politics of hydropower; they are central to how villagers understand what is at stake. The river is not merely a resource to be developed; it is a being with spiritual presence, inhabited by powerful entities whose wellbeing is connected to human wellbeing.

These cosmological voices are almost entirely absent from hydropower dialogues. The government and investors treat the Mekong as a river without beliefs and culture as important elements. The PNPCA process has no mechanism for registering spiritual concerns; EIAs do not assess impacts on naga habitat. This exclusion is not merely an oversight; it reflects the ontological assumptions of modern governance, which recognises only material entities amenable to scientific measurement.

Gendered voices: Women's knowledge and labour

Women's voices are significant and cannot be overlooked in decisions about Mekong hydropower, yet they are frequently silenced in decision-making processes. This silencing operates not only vertically—from state and corporate actors over communities—but also laterally, within communities themselves, where gender, household status, and access to male labour shape who is heard and who bears costs. Women play strong and important roles in households, communities, and cross-community networks throughout the Mekong Basin. They possess detailed and systematic knowledge about the Mekong River ecosystem—fish, plants, aquatic animals—that is as extensive as men's, though often focused on different species and locations (Delfau and Yeophantong, 2020). Women use this knowledge strategically: for making a living, finding opportunities to increase income, providing food for households, and conserving the Mekong River.

The fluctuations and uncertainties in the Mekong's flow caused by upstream dams directly affect women's lives and identities. Earlier research with Bru women affected by Nam Theun 2 in Laos reveals the gendered dimensions of displacement (Manorom, 2017). One widow's story is illustrative:

Before Nam Theun 2 Dam was constructed, her family—her husband and three sons—fished from the Sebang Fai River, grew rice, raised cattle, and collected non-timber forest products from the forests adjacent to the river. Her family lived a life that relied heavily on nature and emphasised sufficiency. Her village was declared at risk of flooding throughout the year. The government ordered all villagers to move to higher ground approximately one kilometre from the village. Before the evacuation, her husband had died, causing her to be unable to evacuate in time with her neighbours because labour was required for the evacuation and building of a new house. The government did not provide a support budget for moving the houses, and only allocated a small piece of land to the affected people.

Her new house was smaller than others—just a house on stilts with a zinc roof and wood. Since her rice field was completely flooded, she allowed her young son to do second-season rice farming with a relative in a distant village to have rice for household consumption. Her eldest son went to work in Bangkok. She now lives with her youngest son, who attends primary school, raising two cows and growing vegetables in a kitchen garden. As she told the researcher: Nam Theun 2 Dam caused her to be separated from her children and have to endure more severe hardship than before.

This account reveals how dam impacts are distributed unequally along gender lines. The widow's vulnerability—the death of her husband just before evacuation, her lack of labour to move her house, the inadequacy of compensation, the separation from her children—reflects the intersection of gender, poverty, and displacement that characterises hydropower's human costs.

Mechanisms of silencing: How voices are excluded

The voices presented above—affective, sonic, corporeal, ecological, cosmological, gendered—are articulate and powerful. Villagers are not silent; they speak with eloquence and precision about what the Mekong means to them and what they stand to lose. Yet these voices have had little influence on the trajectory of Mekong hydropower development. How are they silenced?

First, technical discourse operates as hegemonic. The language of hydropower governance is the language of megawatts and gigawatt-hours, of cost–benefit analysis and environmental management plans. This language is not neutral; it encodes particular assumptions about what counts as knowledge and what counts as value. When villagers speak of the river as kin, of its happiness and suffering, of the nagas who created it, they are speaking a language that technical discourse cannot (or does not want to) hear. Their knowledge becomes, at best, 'local perspectives' to be 'incorporated' into assessments—translated into the dominant language and thereby stripped of its distinctive claims.

Second, the narratives of 'clean energy' and 'development' frame opposition as backward. The state voice in Mekong development proclaims: 'dams are development'; 'dams are clean energy'; 'dams are a tool for reducing poverty'; 'dams can solve flood and drought problems'. These narratives are powerful because they resonate with widely shared values—who could oppose development, poverty reduction, or clean energy? Villagers who oppose dams appear to be opposing progress itself, defending parochial interests against the national good. The fact that Thailand already has a 40 per cent electricity surplus, that dam-related debt is pushing Laos toward economic crisis, that the promised benefits have repeatedly failed to materialise—these facts struggle to gain traction against narratives so deeply embedded in the imagination of modernity.

Third, consultation operates as ritual performance rather than genuine dialogue. The PNPCA process creates the appearance of participation while predetermining outcomes. Power Purchase Agreements are signed before consultations conclude; construction proceeds despite objections. The government listens to the subaltern voices of the villagers and the ecosystem only occasionally and with little attention given to them, or they listen to the local people but do not understand the gist of what the villagers are saying and what issues they are talking about. The state uses its own 'authentic voices' as the norm, relying on core scientific knowledge rather than villagers' situated knowledge.

Fourth, the problem of translation across ontological registers means that even when villagers are heard, their claims are transformed. When villagers say that the river is a person, this becomes 'cultural beliefs about the river'. When they speak of nagas, this becomes 'local mythology'. When they describe the river's happiness, this becomes 'emotional attachment'. In each case, ontological claims—claims about what the river is—are reduced to epistemological claims about what villagers believe about the river. The difference matters: beliefs can be noted and set aside; ontological claims demand a response.

Towards deep listening: Rivers as rights holders

If the problem is that certain voices are systematically excluded from Mekong governance, what would it mean to create conditions for genuine voicing? We propose 'deep listening' as both methodology and ethic—a practice of attending to voices in multiple registers without reducing them to a single metric of value.

Deep listening, as we understand it, requires: attending to what is said (the content of speech); attending to how it is said (the sonic and affective dimensions of voice); attending to who is speaking (the situated position of the speaker); and attending to what cannot be said within existing frameworks (the limits of audibility). Deep listening is not passive reception; it is active work of creating conditions under which diverse voices can be heard.

Emerging legal frameworks offer precedents for institutionalising deep listening by recognising rivers as entities with rights. In Aotearoa New Zealand, the Te Awa Tupua (Whanganui River Claims Settlement) Act 2017 recognised the Whanganui River as "an indivisible and living whole, comprising the Whanganui River from the mountains to the sea, incorporating all its physical and metaphysical elements". The river was declared a legal person with 'all the rights, duties, and liabilities of a legal person'. Te Pou Tupua—comprising one government representative and one Māori representative—was established as "the human face of Te Awa Tupua", charged with speaking for the river in legal and administrative matters (Te Awa Tupua Act, 2017).

The Whanganui settlement was grounded in Māori understandings of the river as an ancestor: 'Ko au te awa, ko te awa ko au'—'I am the river, and the river is me'. As lead negotiator Gerrard Albert explained:

> Rather than us being masters of the natural world, we are part of it. We want to live like that as our starting point. And that is not an anti-development, or anti-economic use of the river but to begin with the view that it is a living being, and then consider its future from that central belief (quoted in Cheater, 2018).

This is precisely the ontological reframing that Mekong governance requires: beginning from the premise that the river is a living entity with its own interests, not merely a resource to be exploited.

In Colombia, the Constitutional Court's 2016 ruling on the Atrato River took a similarly expansive approach. Responding to the devastating impacts of illegal mining on Afro-Colombian and Indigenous communities, the Court declared the river a 'subject of rights' entitled to 'protection, conservation, maintenance, and restoration'. The Court grounded this recognition in what it termed 'biocultural rights'—the inseparable connection between the wellbeing of ecosystems and the cultural survival of communities that depend upon them (Wesche, 2021). The Court ordered the establishment of a Guardian Commission, comprising government representatives and community members, to represent the river's interests—a model that has since been extended to other Colombian rivers including the Magdalena, Cauca, and Coello (Constitutional Court of Colombia, 2016).

Ecuador's 2008 Constitution went further still, enshrining rights of nature at the constitutional level. Articles 71–74 recognise that 'Nature or Pachamama' has the right to "integral respect for its existence and for the maintenance and regeneration of its life cycles, structure, functions and evolutionary processes". Any person can invoke these rights on nature's behalf (Berros, 2015). The Ecuadorian framework draws explicitly on Indigenous concepts—Pachamama (Mother Earth) and sumak kawsay (good living) —to articulate a vision of human–nature relations fundamentally different from the extractive logic that dominates development discourse.

These legal innovations are not without problems. Implementation has been uneven; powerful economic interests continue to override environmental protections; the translation of Indigenous concepts into Western legal frameworks inevitably distorts them (Macpherson et al., 2024). Yet they demonstrate that alternative frameworks are possible—that the ontological assumptions underlying current governance are not fixed but open to transformation.

Applied to the Mekong, a rights-of-river approach would require recognising the river as a stakeholder in decisions about its future—not merely as a resource to be managed, but as an entity with interests that must be represented. This would mean:

- Establishing representative bodies—perhaps modelled on Te Pou Tupua or Colombia's Guardian Commission—comprising government officials, community members, and civil society representatives charged with speaking for the river's interests.
- Recognising villagers' situated knowledge as legitimate expertise—not merely 'local perspectives' to be incorporated, but authoritative claims about what the river is and needs.
- Creating deliberative spaces in which multiple registers of voice—scientific, affective, ecological, cosmological—can be heard and weighed against one another.

- Suspending the presumption that hydropower development is inherently good, and genuinely asking whether proposed dams serve the interests of the river and its communities.

Such an approach would not guarantee different outcomes—the political and economic forces driving Mekong hydropower are formidable—but it would create conditions under which different outcomes become possible. It would shift the burden of justification from those opposing dams to those proposing them, requiring developers to demonstrate that their projects genuinely serve the collective good rather than assuming this from the outset.

Conclusion

The voices of the Mekong—human and nonhuman alike—are vital to decisions about the river's future. Villagers along the Mekong speak with eloquence and precision about their relationship with the river: its happiness and suffering, its sounds and silences, its body and limbs, its fish and spirits. These are not naive or sentimental attachments; they are sophisticated articulations of relational ontology that any Latourian would recognise. The river, in villagers' accounts, is not a resource to be exploited but a being to be respected—a sibling, a friend, a body that can be healthy or sick, a home to spirits whose wellbeing is connected to human wellbeing.

Yet these voices go largely unheard in Mekong hydropower governance. The PNPCA process proceeds; Power Purchase Agreements are signed; construction begins. The discourse of 'clean energy' and 'development' frames dam building as inherently progressive, casting opponents as defenders of backwardness. The framing of hydropower as low-carbon energy lends this discourse particular force in an era of climate concern, yet it obscures the social and ecological costs that communities along the Mekong experience directly. Technical language operates as hegemonic, translating villagers' claims into terms that strip them of their ontological force. Women's voices, in particular, are silenced in decision-making processes, even as women bear distinctive burdens of displacement and impoverishment.

We have argued for 'deep listening' as an alternative approach—a practice of attending to voices in multiple registers without reducing them to a single metric of value. Deep listening would require reconceiving consultation as genuine dialogue across difference, creating space for multiple ontologies of voice, and recognising the river itself as a stakeholder with interests that must be represented. In practice, this might mean reforming the PNPCA process to require that consultations occur before Power Purchase Agreements are signed, that communities have genuine power to refuse projects rather than merely negotiate compensation, and that criteria for project approval include relational and cosmological considerations alongside technical assessments. Legal precedents from Aotearoa New Zealand, Colombia, and Ecuador demonstrate that such recognition is possible, even if implementation remains challenging.

As one villager put it, speaking of the proposed Ban Kum Dam: "We will be able to make money comfortably because we already have our own money in the river. We won't have to struggle or need any help from the government." This is not a plea for compensation or mitigation; it is an assertion of self-sufficiency grounded in relationship with a flowing river. The villager is saying: we do not need your development; we need our river to remain free. Whether this voice will be heard—whether the Mekong will continue to flow as a living river rather than a series of reservoirs—remains to be seen. But deep listening, at minimum, would ensure that the question is genuinely asked.

References

Baran, E., Guerin, E. and Nasielski, J. 2015. Fish, sediment and dams in the Mekong. Penang: WorldFish and CGIAR Research Program on Water, Land and Ecosystems, http://pubs.iclarm.net/resource_centre/Fish-sediment-and-dams-in-the-mekong.pdf (accessed 23 June 2025).

Berros, M.V. 2015. The constitution of the Republic of Ecuador: Pachamama has rights. Environment & Society Portal, Arcadia No. 11, Rachel Carson Center for Environment and Society, https://doi.org/10.5282/rcc/7131 (accessed 20 December 2025).

Cheater, D. 2018. I am the river, and the river is me: legal personhood and emerging rights of nature. West Coast Environmental Law 22 March 2018, https://www.wcel.org/blog/i-am-river-and-river-me-legal-personhood-and-emerging-rights-nature (accessed: 26 December 2025).

Constitutional Court of Colombia (2016) Judgment T-622/16 (The Atrato River Case). Translated by the Dignity Rights Project, Widener University Delaware Law School, http://files.harmonywithnatureun.org/uploads/upload838.pdf (accessed 13 November 2025).

Dore, J., Lebel, L. and Molle, F. 2012. A framework for analysing transboundary water governance complexes, illustrated in the Mekong Region. *Journal of Hydrology* 466–467: 23–36, https://doi.org/10.1016/j.jhydrol.2012.07.023.

Delfau, K. and Yeophantong, P. 2020. State of knowledge: women and rivers in the Mekong Region. Oakland (CA): International Rivers, https://www.internationalrivers.org/wp-content/uploads/sites/86/2020/07/Full-Report-SOK-Women-and-Rivers.pdf (accessed 12 November 2025).

Eidsheim, N.S. and Meizel, K. (eds.). 2019. *The Oxford handbook of voice studies*. Oxford: Oxford University Press.

Haraway, D. 1997. *Modest_Witness@Second_Millennium.FemaleMan_Meets_OncoMouse: feminism and technoscience*. London: Routledge.

Herzfeld, M. 2004. *Cultural intimacy: social poetics in the nation-state*. 2nd edn. New York (NY): Routledge.

Johnson, A. 2020. *Mekong dreaming: life and death along a changing river*. Durham (NC): Duke University Press.

Kanngieser, A. 2012. A sonic geography of voice: towards an affective politics. *Progress in Human Geography*, 36(3): 336–353, https://doi.org/10.1177/0309132511423969.

Latour, B. 1993. *We have never been modern*. Translated by C. Porter. Cambridge (MA): Harvard University Press.

Latour, B. 2004. *Politics of nature: how to bring the sciences into democracy*. Translated by C. Porter. Cambridge (MA): Harvard University Press.

Macpherson, E., Borchgrevink, A. and Cribb, M. 2024. Beyond legal personhood for the Whanganui River: collaboration and pluralism in implementing the *Te Awa Tupua Act*. *The International Journal of Human Rights*, 1–24, https://doi.org/10.1080/13642987.2024.2314532.

Manorom, K. 2017. Lao women in Xe Bang Fai Basin: indigenous people, intersectionality, emotional geographies and impacts of Nam Theun 2 Dam. *Journal of Development Studies*, Puey Ungphakorn School of Development Studies, Thammasat University, 13(2), 1–45, https://s005.tci-thaijo.org/index.php/gvc-tu/article/view/95621 (accessed 29 March 2026).

Manorom, K. 2022. *Mekong River ontology: dams, water and people*. Bangkok: Siam Publishing House.

MRC (Mekong River Commission). 2010. Proposed Mekong mainstream dams. Basin Development Strategy Background Paper. Vientiane: MRC Secretariat.

Permphongsacharoen, W. 2018. Thai electric power remains sufficient. [in Thai]. Bangkok: Foundation for Ecological Recovery.

Revill, G. 2016. How is space made in sound? Spatial mediation, critical phenomenology and the political agency of sound. *Progress in Human Geography* 40(2): 240–256, https://doi.org/10.1177/0309132515572271.

Revill, G. 2021. Voicing the environment: Latour, Peirce and an expanded politics. *Environment and Planning D: Society and Space* 39(1): 121–138, https://doi.org/10.1177/0263775820944521.

Schmitt, R.J.P., Bizzi, S., Opperman, J. and Kondolf, G.M. 2019. Planning dam portfolios for low sediment trapping shows limits for sustainable hydropower in the Mekong. Science Advances 5(10): eaaw2175, https://doi.org/10.1126/sciadv.aaw2175.

Sen, A. 1981. *Poverty and famines: an essay on entitlement and deprivation*. Oxford: Clarendon Press.

Soukhaphon, A., Baird, I.G. and Hogan, Z.S. 2021. The impacts of hydropower dams in the Mekong River Basin: a review. Water, 13(3): 265, https://doi.org/10.3390/w13030265.

Spivak, G.C. 1988. Can the subaltern speak? In Nelson, C. and Grossberg, L. (eds.), *Marxism and the interpretation of culture*. Urbana (IL): University of Illinois Press, 271–313.

Te Awa Tupua (Whanganui River Claims Settlement) Act 2017. New Zealand Public Act 2017 No. 7, https://www.legislation.govt.nz/act/public/2017/0007/latest/whole.html (accessed 30 October 2025).

Wesche, P. 2021. Rights of nature in practice: a case study on the impacts of the Colombian Atrato River decision. *Journal of Environmental Law* 33(3): 531–555, https://doi.org/10.1093/jel/eqab021.

Wongthet, S. 2023. Mekong River nagas connect the Thai-Lao and Mon-Khmer kinship. *Arts and Culture Journal* 11 June 2023, https://www.silpa-mag.com/history/article_95580?utm_source=chatgpt.com (accessed 29 March 2026).

Yvonne, G.A. 2017. Potent places and animism in Southeast Asia. *The Asia Pacific Journal of Anthropology* 18(5): 389–399, https://doi.org/10.1080/14442213.2017.1401324.

CHAPTER 10

Biophysical geopolitics and hydrosocial relations in the environmental geopolitics of the Mekong Basin

Carl Grundy-Warr[1]

Abstract

Basin-scale environmental geopolitics is dominated by debates about 'sustainable hydropower' and big water projects. Dominant narratives stress regional and national economic benefits, whilst counter-narratives focus on cumulative negative social and ecological impacts. This chapter proposes that within broader environmental debates, there are critical biophysical, nonhuman, and hydrosocial matters operating at transboundary, subnational scales, positions, and places. The terms 'biophysical geopolitics' and 'hydrosocial relations' are used to discuss case study vignettes that stress the need for policymakers to go beyond a one-dimensional focus on megaprojects and 'sustainable hydropower' that override protection of vital ecological, cultural, and nonhuman resources at subnational transboundary scales and across many positions and places within the basin.

Keywords: Biophysical geopolitics; hydrosocial relations; environmental geopolitics; transboundary subnational scale, position and place

Introduction

Political economic and developmental decisions regarding environmental resources necessarily involve transforming biophysical properties and processes. In the Mekong Basin, there are massive transformations to the physical landscape through large numbers of megaprojects working relentlessly through ecosystem processes, that serve to reshape and disturb them (Harvey, 1996), and creating 'anthropogenic rivers' (Whitington, 2018). This has profound implications for the riparian and wetland communities that share the basin who have little alternative but to cope and adapt to the socio-environmental impacts and cumulative consequences of these interventions (Soukhaphon et al., 2021).

[1] Department of Geography, National University of Singapore.

Biophysical geopolitics can be viewed as a subset of environmental geopolitics that focuses on parts of the 'bio' (living nonhuman nature) and 'physical' landscape (deep pools, rocks, reefs, channels, geomorphological features, sediments), which relate to those parts of the basin that are affected by these megaprojects, such as navigation schemes, water diversion projects, and hydropower dams. Large-scale technologies and infrastructure inevitably have implications for the politics of scale, position, and place. I apply the notion of b\ordering to suggest that hydropower and other infrastructure have different kinds of geographical bordering and social reordering implications (Grundy-Warr, 2017). Hydropower plants create semi-enclosures—privatised corporate spaces—with their own strategic inclusions (for company personnel and official visitors) and exclusions (of everybody unrelated to their operation). They reconfigure socio-environmental and hydrosocial relations by fragmenting and reordering ecosystems to create "new techno-borders of space and resources, which transform bits of former commons into privatized fiefdoms" (Yong et al., 2024: 290). Cascades of dams on tributary systems completely alter whole sets of relationships between communities of people and their environment, often leading to displacement and hydrosocial rupture (Chann et al., 2024).

This chapter starts by providing an overview of regional eco-social-political arguments reflecting on the tensions between its development ambitions and sustainability. It then goes on to present a series of vignettes that help to illustrate why flexible focal lengths are helpful to show regional and transborder environmental geopolitics whilst having a shallow depth of field that enables a focus on particular biophysical objects and issues. The impacts of megaprojects are never uniform, but have variegated, complex, and diverse implications across time and space. A regional-scale environmental geopolitical perspective can help challenge assumptions about what is 'sustainable' about developments in the whole Mekong Basin, and a more finely tuned biophysical and hydrosocial focus can explore specific issues that differ according to 'scale, position, and place' within complex hydrological basins and along river systems (Lebel et al., 2005). For each vignette, there are scale, position, and place considerations, which generate questions about water governance and environmental geopolitics in the Mekong Region.

Geopolitics of a transformed river basin

Particular reefs, channels, deep pools for dolphins and fish, and other biophysical matter, have become geopolitical at different moments and phases of development in the Mekong Basin. It is possible to talk about geopolitical rocks, pools, channels, and nonhuman living matter when they become integral to developmental narratives that straddle and are impactful at various scales, regional, transboundary, and national, all the way down to subdistrict and village scales. Grounded geopolitical river issues require unpacking, not least because socio-ecological relationships in the Mekong Basin are simultaneously regional, transboundary, trans-local yet intensely localised in terms of ecological, economic, social, and cultural impacts. Environmental geopolitics can help to analyse basin-scale environmental and social impacts and raise questions regarding 'sustainable development'. A focus on specific biophysical geopolitics reveals why transboundary subnational, fine-tuned scales, multiple locations, positions and places are critical to seeing diverse and differentiated impacts of megaprojects.

The concept of 'the hydrosocial cycle' incorporates the hydrological knowledge of the water cycle with the intimate interactions between water and people, that are not just material, but also "experiential, cultural and metaphorical" (Linton and Budds, 2014: 174). The hydrosocial cycle is an analytical approach that integrates the hydrological and social aspects of the water cycle together, focusing on the intimate interactions between people and water. It relates to the discussion about biophysical geopolitics because it sees that capital-socio-natural relations modify aspects of the hydrological cycle. Seeing "the circulation of water – as a biophysical and social process – brings to light wider political economic, social, and ecological processes" (Swyngedouw, 2004: 2). In this chapter, each of the vignettes illustrates how biophysical geopolitics have hydrosocial and nonhuman ecological dimensions that are of relevance to critical questions concerning larger developmental and environmental discourses and practices at basin-wide and regional scales.

In the space of three decades, the Mekong River Basin has been transformed from an almost free-flowing 'under-developed' river basin (Sluiter, 1992) to an anthropogenic geo-engineered one (Grundy-Warr and Middleton, 2025) whereby "biophysical processes and entities" are "incorporated into capitalist orbits of production and circulation" (Sneddon and Fox, 2012: 281). The way that we consider hydropolitics has changed over time, altering nodes, networks, and assemblages, particularly as the river itself is cumulatively reproduced, reshaped by hydropower projects. At a global and sub-regional scale, it is relevant to consider how, over time, dominant developmental and environmental discourses have generated changes throughout the Mekong Sub-Region, and within the hydrological basin.

Understanding discursive aspects of geopolitics enables us to observe the development of powerful techno-political assemblages and discourse coalitions across and within state borders. A geopolitics lens helps to see how dominant state-centric discourse may generate and perpetuate certain simplifications of geographical reality through ideas such as 'modernisation', 'common prosperity', and 'sustainable development'. All of these ideas have a tendency to homogenise space and nature and to work in specific ways that transform and close off other potentials.

One aim of this chapter is to highlight the discursive ways in which biophysical properties become part of geopolitical imaginaries or regional, national, and 'sustainable' development. Nonhuman nature is an active agent in developmental and political ecological matters. Actor–Network Theory highlights the ways in which humans are part of complex networks and assemblages including nonhuman and material elements (Latour, 2004, 2005; Law, 2009). This contributes to Mekong scholarship that stresses the fundamental importance of biophysical diversity and the threats that homogenising forms of development pose to future ecological (and social) wellbeing (Baird and Hogan, 2023). It relates to a neglect of centring ecology and biodiversity within development.

In the Mekong, the waterscapes and associated biophysical features are active elements in efforts to transform, commodify, and control them. The nonhuman elements of anthropogenic projects (dredging, canals, dykes, dams, boats, dynamite) have "material powers, which can aid or destroy, enrich or disable, ennoble or degrade us" (Bennett, 2010: ix). The environmental history of the Mekong Delta reveals the interplay of the nonhuman, physical, technological, and political in efforts to mould the delta frontier into pre-colonial, French colonial, American War, and nation-building projects. "Because of these multiple, overlapping layers of constructed spaces and the delta's amorphous, amphibious nature, it serves as an ideal environment for a history of modernization oriented to physical examples of slippage, erasure, and rupture" (Biggs, 2010: 6). Elsewhere within the Mekong Basin, there are multiple ways in which riverscapes, and landscapes, have been shaped and reshaped by water resource development projects. This blending of social and physical space reflects the 'hydrosocial cycles' introduced above.

With an intensification of hydropower and other megaprojects, there is not only a need to rethink meanings of 'nature' within geopolitics and development (Grundy-Warr et al., 2015), but to re-position (elevate) the standing of biophysical processes, bits of nature, and hydrosocial relations within ideas about regional and national development. Water (and energy) governance should not ignore the 'unseen' transboundary commons of the Mekong (Grundy-Warr and Lin, 2020), or overlook "the biological resources being lost" (Campbell and Barlow, 2020), just as there is a vital need for cumulative transboundary impacts to be reckoned with and hidden costs of lost ecological services to be seriously considered (Soukhaphon et al., 2021). Hydrological projects have led to new forms of b/ordering by physically enclosing and privatising former commons (Grundy-Warr, 2017), and by re-ordering 'what counts as nature' (Castree and Braun, 2001).

Over time, cumulative projects alter natural flow patterns, rhythms, and cycles of river basins. In doing so, "important ecosystems fail, often resulting in failure of human life-support systems" with deleterious consequences for wetlands and river cultures (Wantzen, 2022: 1). As Baird and Hogan (2023: 8/21) observe: "river fragmentation from dams increases extinction risk" and creates ecological "extinction debt" between the periods of environmental impact and species disappearance. Whilst some governments in the region may recognise the need for limited degrees of conservation or environmental protection, dominant developmental and environmental narratives have become drivers of political economic actions that ultimately lead toward greater ecological fragmentation within and across borders. The idea that massive basin-wide hydropower development is 'sustainable' is backed by powerful coalitions of state, financial, and corporate decision makers that militate against the sorts of ecological flows and environmental connectivity advocated by concerned scientists aware of the multiple, less sustainable, dimensions of the industry.

Campbell (2009: xiii) has noted that

> [d]ecisions about the future of the Mekong, are ultimately political and must be made by the people of the region. However, those decisions should be informed as much as possible by an understanding of the biophysical nature of the system and the likely consequences of various development trajectories on the river and its ecosystems, and ultimately on themselves.

A number of salient issues are contained in this statement about the politics of nature and environment, and about the politics of knowledge. Key decisions are 'ultimately political' making certain choices that then discard alternatives (Geheb and Suhardiman, 2019). These decisions are made by decision makers in elite institutions and ministries with considerable power and authority. Whilst Campbell's (2009) emphasis is on biophysical matters and the relevance of scientific approaches, it is equally necessary to stress that all ecological projects are socio-political ones that are driven by the developmental agendas and ambitions of regional political elites. The extent to which 'the people' are involved in the developmental decisions that have transformed the resource economy and socio-natural landscape of the Mekong is questionable. Campbell's call for 'people of the region' to play a part in these decisions assumes levels of participatory deliberative politics that have typically been restricted in regional scale and scope (Dore and Lebel, 2010; Geheb and Suhardiman, 2019; Suhardiman and Geheb, 2021; Yong and Gillespie, 2022).

The politics of knowledge relates to how knowledge(s) inform and shape actual policymaking and decisions in the Mekong River Basin. Over the years, the foremost transboundary mechanism for producing knowledge and information has been the Mekong River Commission (MRC). Anybody who has visited the MRC Secretariat library will see that it is stocked with excellent information, scientific studies, and technical reports about many of the biophysical attributes of the Mekong hydrological basin. There is a wealth of science to draw upon to inform decisions about sustainable development that take into full account issues of environmental sustainability and the needs of the basin's 'people'. These publications and many of the technical reports do not inform policymaking at the higher level. The developmental emphasis has been on large hydraulic infrastructure, in amongst multiple big commercial land commodification projects, and plantations that are rapidly transforming water and terrestrial environments of the region (Hirsch et al., 2022).

Geopolitical discourse and sub-regional development

Ó Tuathail and Agnew (1992: 193) argue that

> discourses enable. One can view these capabilities or rules as permitting a certain bounded field of possibilities and reasoning as the process by which certain possibilities are actualized. The various actualizations of possibilities have consequences for the further reproduction and transformation of discourse. The actualization of one possibility closes off previously existent possibilities and simultaneously opens up a new series of somewhat different possibilities.

While a powerful discourse directed by political elites may close off other possibilities, this does not mean that the discourse is static and unchanging. Discourse is part of geopolitical reasoning which "works by the active suppression of the complex geographical reality of places in favour of controllable geopolitical abstractions" (Ó Tuathail and Agnew, 1992: 195).

Taking an 'environmental geopolitics approach' is to examine "how environmental themes are used to support geopolitical arguments and realities. It asks how the environment is brought into narratives, practices, and physical realities of place and power" (O'Lear, 2018: 2). Views about 'the environment' will be partly shaped by a stakeholder's position in relation to any given project, which brings into question the nature of technical politics and roles of engineering consultancy (Boer et al., 2016). Castree (2003: 427) discusses how geopolitical environmental discourses become shaped by powerful political economic alliances that shape how environmental and sustainability questions are framed: "The framing process works through a process of discursive inclusion / exclusion, as some things are deemed 'legitimate' subjects while others are simply off the agenda." Coalitions of powerful institutional, state and corporate actors seek to monopolise this agenda setting, and to steer the trajectory of the discourse.

Close links between practical geopolitical reasoning, hydropower project promotion, and regional developmental discourses have been evident in the Mekong since the 1950s (Sneddon, 2015), emerging from the Cold War activities of the US

Bureau of Reclamation and echoed through the plans of the United Nations supported Committee for the Coordination of Investigation of the Lower Mekong Basin ('the Mekong Committee') established in 1957, the Cold War forerunner of the MRC. These US-influenced institutions during the Cold War were instrumental in shaping the idea of regional geopolitical pacification via engineering, technology, and big infrastructure. These initiatives incepted a narrative around the fundamental need for large-scale infrastructure if grand visions of modernity and development were to be achieved (Wong, 2010).

The 1995 signing of the Lower Mekong Agreement between Thailand, Laos, Cambodia, and Vietnam signalled "the new regional geopolitics of rapprochement" (Hirsch, 2016: 67) and led to the formation of the MRC. Operating independently outside the 1995 Mekong Agreement, China has its own imagined geographies of the Lancang-Mekong. The end of the Cold War saw the unilateral development of a cascade of dams along the upper Mekong mainstream. Yunnan Province emerged as a 'gateway' to the rest of Southeast Asia, with hydropower as one of the means to gain a stronger presence as an investor, financier, and developer within the region (Glassman, 2010). "The PRC wants to electrify Southeast Asia and it is 'going out' with a low China price to do it – for profits, for resource control, for regional influence" (Grumbine, 2010: 113). Hydropower dam technology and technical assistance were conceived of as a 'tool of influence' based upon China's global hydropower technical expertise and political economic leverage as a key investor in the region (Sneddon, 2015: 139). Chinese state-owned enterprises (SOEs) and financial institutions became integral to the subsequent expansion of hydropower in Southeast Asia as a 'driver' of regional development (Grumbine et al., 2012).

The creation of the Lancang-Mekong Cooperation (LMC) framework in 2015 enabled China to push its own discourse of regional cooperation based on norms-consensus and bilateral ties developed through various infrastructure projects, including hydropower (Grünwald, 2021; Grünwald et al., 2022). The LMC casts itself as a part of the Belt and Road Initiative (BRI) but focused on sub-regional cooperation, cross-border projects, and strong bilateral ties. It has been described as "a prime example of Sino-centric multilateralism, in which China is the one who sets the rules and frameworks rather than being absorbed into pre-existing institutions such as the [MRC]" (Hoang, 2022: 2). The LMC is not focused on the sustainable utilisation of water, but on multiple projects aimed at improving connectivity through infrastructure, technology, and cross-border development, ranging from agriculture to transport to public health.

China has sought to promote 'common development goals' through numerous engagements that downplay its dam-building projects (Biba, 2012). "Through discourse, Beijing aims to build a positive image of itself by framing its achievements, experiences, and ideas as ways to enhance regional cooperation" through narratives of "equality, local autonomy, non-interference, and reciprocity" (Gong 2022: 24). The LMC has rapidly become a powerful mechanism covering the whole Mekong Sub-Region, with greater political economic influence over state elites than the MRC. Transboundary infrastructure projects help strengthen diplomatic ties with Beijing, and they dovetail with the political economic agendas of national elites. Thus, the MRC has become a mechanism for promoting 'sustainable hydropower' legitimised by compliant technical assessments and engineered mitigation of environmental harms. There is synergy between the 'sustainable development' promoted by the MRC and the ongoing cross-border infrastructures endorsed through the LMC. "Dams, as material expressions of development, depend on the state embracing and elevating the technological and engineering knowledge that underpins their construction and operation" (Fox and Sneddon, 2019: 4, 18). Linking ideas about 'sustainable development' with regional and national development have enabled top-down state–capital controls that effectively close off other pathways.

At regional and national levels of analysis, discursive geopolitics may be linked to the way in which states, financial institutions, and capital interests have promoted ideas about 'sustainable hydropower' as 'sustainable development' based on geographical imaginaries of river basin development. The MRC has become a prime promoter of 'strategic priorities' and 'sustainable development goals' through its 10-year basin development strategy (MRC, 2021). Creating dominant developmental and environmental narratives requires the selection of certain types of 'expert' knowledge which may be used selectively to justify and legitimise ecological or developmental projects. In turn, these become "devices through which actors are positioned, and through which specific ideas of 'blame' and 'responsibility' and 'urgency' and 'responsible behaviour' are attributed" (Hajer, 1995: 64–65). As Geheb et al. (2015: 101) argue in their detailed analysis of the use of environmental narratives in Laos, "narratives surrounding hydropower often delineate the spaces between winners and losers of development". Narratives often draw strategically upon selective bits of technical and scientific knowledge as a way to legitimise certain developmental projects over others, and in turn, such selective applications of knowledge help maintain and shape the social order in favour of the power holders (Jasanoff, 2012). "As a result, environmental narratives frequently impose meanings that are acceptable to their creators and users, but which may contain unwelcome implications for other social actors and high levels of simplification of complex and unwelcome physical processes" (Forsyth and Walker, 2008: 17).

Environmental narratives help create 'problem closure' (Hajer, 1995) whereby certain explanations, coupled with applications of apparently politically neutral scientific and technical expertise, serve to lessen debate by obscuring counter-narratives from entering into the public domain. This does not mean that counter-environmental narratives do not exist. Rather, the environmental narratives work as devices that may shape, influence, and nullify those counter-narratives, aided by restricting who participates or is consulted. As Yong and Gillespie (2022: 11) observe: "multi-scalar and heterogenous state authorities may influence forms and degrees of participation against hydropower development." Powerful actors may change power and authority by working at different spatial scales, which may influence decision-making processes, including approved participants and excluding troublesome others from those processes (Swyngedouw, 1997).

The creation of dominant developmental and environmental narratives centring on hydropower became a means to legitimise policy decisions whilst stifling opposing or alternative ideas about development. The Mekong Water Resources Assistance Strategy (World Bank and ADB, 2006), for example, while pushing for 'balanced development' and 'sustainability', "... unashamedly privileges hydropower investment and water development at the expense of alternative uses of water, or other methods of generating electricity" (Geheb et al., 2015: 105). As Sarkkula et al. (2009: 243) have noted: "[b]asin developments are repeatedly and without critical views being justified by the statement that large-scale infrastructure construction is the solution to reducing poverty, although there is no evidence this will lead to the claimed result." Hydropower is promoted as both 'clean green' energy and as a motor of economic development. Yet this overlooks uneven development outcomes, for electric power flows mostly to urban centres and across borders. The vast majority of Lower Mekong rural dwellers are "involved in water resource related activities as their principal occupation" and hydropower "re-allocates the benefits of water use away from the local level" (Williams, 2020: 12884).

In the Mekong, the 'tangled nets of discourse' relate to environmental narratives officially applied in support of hydropower as opposed to freshwater fisheries (Yong and Grundy-Warr, 2012). In the region, state elites have tended to portray 'wild capture' fisheries as a symptom and even cause of poverty, suggesting that over-fishing by poor fishers is responsible for fisheries decline, and not hydraulic infrastructure (Friend, 2009). The poverty reduction claims by hydropower proponents regarding riparian and wetland communities are a myth if livelihood alternatives are not found to replace the loss of ecosystem services emerging from hydropower's cumulative impacts (Baran and Myschowoda, 2009; Ziv et al., 2012). Constructing counter-narratives is relevant to the task of bringing different ways of utilising water resources and alternative forms of development into policy arenas.

Discursive geopolitics is "contingent on the power relations of particular places and tend to serve the interests of those who stand to benefit economically and politically from those interventions" (Fox and Sneddon, 2019: 6, 18). Thus, the political economy of hydropower motivates and reinforces the continual refinement and reiteration of the regional infrastructure development narrative. "The political utility of the hydropower myth has further compounded the myth's position. Hydropower's symbolic role in enforcing state legitimacy and dominance can be considered by the political elite to justify pernicious trade-offs that result in negative environmental and social externalities" (Williams, 2020: 12887).

Powerful 'discourse coalitions' promoting hydropower are multi-scale, but they profoundly alter the grounded realities for ecosystems and communities at transboundary and subnational scales. These 'discourse coalitions' encompass an array of political elites, energy and water ministries and agencies, international financial institutions, national banks, SOEs, and private firms. State–capital–financial alliances comprise economic beneficiaries, including national energy agencies, for the ecological projects that they actively promote (Geheb and Suhardiman, 2019). Hydropower alliances are forged within already existing global capitalist forces operating in the Mekong sub-region (Glassman, 2010), supported by finance, engineering expertise, and technological knowhow from China (McDonald et al., 2009; Xiao et al., 2023; Zhou et al., 2020).

Volumetric geopolitics

The culmination of the intensive development of hydropower has transformed 'cascading geopolitics' (Hirsch, 2011) into 'powersheds' (Magee, 2006; Middleton and Allouche, 2016) and into basin-wide 'volumetric geopolitics' (Grundy-Warr and Middleton, 2025). The importance of transboundary water data has grown, on the one hand, due to increasingly severe regional droughts and on the other due to the water storage capabilities of the region's dams (Grünwald et al., 2021; Ketelsen et al., 2020a). In spite of the MRC and China signing an agreement on water data sharing in 2002 (renewed in 2008), there continue to be anxieties about upstream water supply to the lower Mekong. These, then, are volumetric concerns.

Unusually low Mekong flows during 2019-2020 were the focus of an Eyes On Earth report using time-series hydrological data and remote sensing, which claimed to demonstrate that upper Mekong dams were withholding water for their reservoirs, thereby contributing to lower Mekong River flows and compounding a severe drought (Basist and Williams, 2020).[2] Subsequently, the US-based Stimson Center launched a 'Mekong Dam Monitor' (MDM) that uses the Eyes On Earth methodology to report water levels and releases from 26 'virtual gauges' located at dam sites on the mainstream and tributaries, rendering their water releases and operation more visible. China's Foreign Ministry welcomed the MDM as a 'constructive action' but warned against 'malicious moves' to 'drive a wedge' between

[2] The methods employed in the study have been critiqued. See, for example, Kallio and Fallon (2020) and Ketelsen et al. (2020b).

China and the Lower Mekong countries (Grünwald, 2021: 9). In October 2020, the MRC announced a 'landmark agreement on China–MRC cooperation' with China agreeing to provide the MRC with year-round hydrological data "contributing to better river monitoring and flood and drought forecasting in the Mekong countries" (MRC, 2020b).

Volumetric geopolitics also relates to how dams that secure flows and impact on basins' hydrological regime have altered the topologies of power beyond territorial sovereignty to more complex techno-political relations across borders (Allen, 2016). As hydropower development has continued apace, its impacts on the Mekong's flows have become increasingly obvious (Hecht et al., 2019). At one level, this has meant that hydropower proponents have turned to demonstrate that if hydropower dam operations were coordinated, then these impacts could be mitigated—in other words, using hydropower to mitigate the impacts of hydropower. Better still, if hydropower were coordinated in these ways, then the dividends could be substantial: dam releases could increase flows during dry seasons or drought, while reducing flows and flooding during wet seasons (Yu et al., 2024). Hydropower, properly coordinated, can help the Mekong countries to cope with the vagaries of climate change. In this light, 'designing flows' takes on new geopolitical meanings with implications across scale, position, and place (Lebel et al., 2005).

"The method for design that we propose leverages a highly flexible and well-understood spectral toolbox that would have direct, transferable applicability to systems engineers who operate dams. In short, the design provides a potential tool for managing mainstem hydrology for fisheries as new dams come on line" (Sabo et al., 2018: 1271). In this manner, "science would provide data-driven models to enable geo-engineering supporting 'sustainable' hydropower" (Grundy-Warr and Middleton, 2025). Whilst there is little doubting the scientific expertise here, there are other branches of science to be considered, including fisheries knowledge. "Mekong fish require more than designed flows", and that "ignoring the power objectives of dam operations" makes it highly unlikely that such hydro-engineered approaches for regulating flows through the hydrological system will be adopted (Hogan et al., 2018: 1). Furthermore, evidence from other forms of engineering mitigation in relation to migratory fishes, such as fish ladders, bypasses, and dredging channels near to dams have questionable efficacy (Campbell and Barlow, 2020; Pukinskis and Geheb, 2012). Riparian communities require more than designed flows too, as aquatic resources and fisheries remain critical issues for livelihood and food security. As Wantzen (2022) observes: "[t]he re-establishment of designer flow regimes alone – e.g., in dam management – is generally not sufficient to re-establish human–river relationships" developed over long periods of time. "Geo-engineered solutions are abstracted from the social lives, wetland cultures, and everyday hydro-social realities in the Mekong, and there is a need to reconsider the intimate social riparian relationships alongside the technological changes being made to biophysical matter."

Biophysical matter and geopolitics

I make a distinction between large-scale environmental geopolitics at global, regional, and inter-state levels, and a focus on biophysical geopolitics that concentrates on distinct aspects of the physical environment or discrete socio-ecological relations or nonhuman actants at subnational transboundary levels of analysis.

Biophysical geopolitics spotlights how specific physical landforms or features and nonhuman living and non-living elements become implicated within controversial elements of ecological projects that are themselves integral to bigger-scale regional and international environmental geopolitical narratives. A biophysical geopolitics lens is not separate from, but a part of a critical environmental geopolitical approach, but one that stresses subnational yet transnational scales often bypassed or underestimated at regional, international, and national policymaking levels. Biophysical relations are entwined with the politics of position and place (Lebel et al., 2005). The politics of position, such as a particular riverbank, a specific tributary, location near an estuary, a migratory fish channel, a deep pool, or being located near a river border, can play consequential parts within localised arenas that reflect upon regional, inter-state, and national environmental geopolitics. The cultural landscape and place matter greatly, as there is a multiplicity of different hydrosocial place-based relations along the rivers, lakes, floodplains, and wetlands of the Mekong Basin. Hydrosocial relations are often place-based associations with specific environmental resources, things, and water bodies. In the lower Mekong, many places are embedded within spaces of dependence that provide food security, livelihoods, and identity to the people who live there. As Escobar (2001: 140) puts it: "we understand by place the experience of a particular location with some measure of groundedness (however, unstable), sense of boundaries (however, permeable), and connection to everyday life, even if its identity is constructed, traversed by power, and never fixed."

In considering nature's lively materiality within geopolitics (Sneddon, 2007), the focus of this section is on biophysical case study vignettes that focus on very specific parts of the Mekong basin where certain bits of nature have become agents within broader issues of transnational environmental politics and geopolitics. These vignettes reveal aspects about the "spatial dimensions of human-environment relationships that occur unevenly in different places and are intertwined with local, political, and cultural geographies" (O'Lear, 2018: 7). Each vignette relates to 'grounded' geopolitical issues relating to mega water projects as they impact upon multiple places, ecological processes, and 'local' environments in the Mekong Sub-Region. For each of the cases, there are overlapping scales to consider, in addition to the peculiar characteristics of in situ politics, position, and place. "The scale (and levels) at which a problem is experienced, analysed, and discussed, or 'scale of meaning,' may not correspond to the scale of the decision-making bodies" (Lebel et al., 2005: n.p.). For each case, there are peculiar in situ cultural and vernacular readings of the physical landscape and nonhuman nature that contrast sharply with state-centred abstractions and technocratic visions of nature.

Geopolitical rocks and rapids

Historically, rocks, reefs, and shoals have often become features that figure prominently in traditional geopolitical issues, such as border-making, territory-making, and sovereignty disputes. A 'geologic turn' (Clark, 2013) raises interesting questions about the geopolitics of the transformation of geological and geomorphological features by corporations and states. Studies beyond the Mekong reveal how modifying bits of the biophysical environment can be geopolitical, such as island building, and alterations to maritime reefs, atolls, and shoals, and using massive amounts of imported material and sand for land reclamation in contested spaces of the South China Sea (Schofield, 2013). Rice et al. (2016: 447) examine 'the biopolitics of geophysical transformations' that have geopolitical significance by altering claims and counter claims to physical features and sovereign rights over natural resources and sea space. In the Mekong Basin, rocks and reefs play a vital role as refuges, feeding grounds, and breeding areas for different fish and bird species, and they are also part of the political landscape, particularly as they are often used in delimiting where riverine boundaries run.

Figure 10.1 is a photo of a section of the Mekong River between Chiang Khong in Chiang Rai Province in northeastern Thailand and Bokeo Province in northwestern Laos, showing one section of the local reefs known as Khon Phi Long, or 'waterfall' (as the water flows over the rapids in the wet season) 'of lost ghosts' or the 'Islet of Lost Souls' (Techawongtham, 2017). According to folklore, in the past, deceased ancestors were floated downstream and the rocks harbour their spirits (Yong, 2020). For local people, boat operators, traders, and fishers, the river unifies more than it divides, and geomorphic features in the river often have meanings and values that relate to beliefs, practices, and customs. In addition to the rocks' ancestral and spiritual significance, they are excellent places to place fish traps (economic) and are 'second homes' for local fishers from both sides of the boundary (cultural transnational linkages). In spite of the river boundary making most river islets part of Laos, local fishers from each side meet on some of the larger rocks and sandy islands.

Figure 10.1: Rocks and reefs of Chiang Khong—Bokeo river border space. Source: Author.

Fisheries science reveals the value of different rocks, reefs, and deep pools as fish habitats and refuges for differing species assemblages linked to a broader ecosystem (MRC, 2015). Due to rising and falling waters in relation to wet and dry seasons, features within the river and the riverbanks have vertical and volumetric dimensions, particularly as exposed features are used for fishing and many riverbank gardens are utilised in the dry season. Hydropower developments contribute to less certainty about seasonal and even daily water levels in some downstream locations. Understandings of socio-ecological, hydrological, and biophysical attributes have broader implications for geopolitical contestations over forms of resource development for they "serve as a basis for innovative political thinking and practice" (Sneddon, 2007: 187).

From 2001-2003, rocks, reefs, and sandbanks in the river became the focus of transnational geopolitics as Rak Chiang Khong ('Love Chiang Khong'), a community-based organisation aided by NGOs through the Southeast Asian Rivers Network, resisted plans to blow up these geomorphological features as part of the China-led Navigation Channel Improvement Project. The latter was part of the Quadripartite Economic Cooperation Plan for the 'navigational improvement' of the Lancang-Mekong from Simao (China) to Luang Prabang (Laos) (Hensengerth, 2009). The geo-economic imaginary of the Mekong as a super highway for vessels up to 300 DWT (Dead Weight Tonnage) was contested by local CBOs such as Rak Chiang Khong, whose activists lobbied the Thai Border Patrol police about their concerns over the effects of blasting on the river boundary with Laos. In July 2002, the Thai Ministry of Defence (MoD), under then-Defence Minister Chavalit Yongchaiyudh, called for a temporary suspension of the project due to concerns that it would impede future river border delimitation and demarcation work (Southeast Asia Rivers Network, 2003). The MoD were concerned that border problems could emerge if the rocks and reefs were removed, and doubts emerged about the precise location of the thalweg, the line or course of the lowest elevation within a watercourse (Osborne, 2019). This boundary was first delineated in an 1893 treaty signed between France and the Siamese government and modified in 1926. The boundary line follows the thalweg nearest to the Thai side, which means that most of the islands belong to Laos (United States Department of State, 1962).

The biophysical geopolitical nature of this dispute brings in multiple political economic and geopolitical scales. Whilst river navigational improvement is not formally a part of the BRI, it fits into the BRI imaginary of an intensified network of rail links, roads, bridges, and port developments that are related to notions of a 'China-Indochina Peninsula Corridor' (Rowedder, 2020). Chinese trading companies operating large vessels up to 500 tonnes downstream would be the primary beneficiaries of blasting and deepening the river. "Chinese vessels are more powerful and more numerous than competing Lao vessels and the trade that takes place is dominated by Chinese firms" (Osborne, 2019). MRC imaginaries of the Mekong as a super highway for trade were invoked in support of the scheme, with the then-MRC Secretariat CEO, Dr. Pham Tuan Phan, publicly arguing for "improving" the river for increased river trade, suggesting that environmental campaigners should support such a move as river trade is less polluting than overland trucks using highways (Pham, 2017). The state-centred geo-economic transboundary and regional scales relate to respective ministries of transport in Laos and Thailand, the MRC, and Chinese river trading companies, producing developmental narratives that are superimposed over alternative livelihood, cultural, and ecological considerations at finer scales, positions, and places along the river.

After the initial success of the Chiang Khong Conservation Group alongside relevant NGOs in stopping the blasting of the Khon Phi Long and other features, the plan was resurrected again in December 2016 by General Prayuth Chan-ocha's military government in response to The Development Plan for International Navigation on the Lancang-Mekong River (2015-2025) (Yong, 2020). As part of the drive to get Thai approval, the principal Chinese multinational company involved, CCCC Second Harbour Consultants, conducted a series of consultation meetings at key districts along the Thai banks of the Mekong, in Chiang Saen, Chiang Khong, and Wiang Kaen, Chiang Rai Province. In these dialogues, rocks, reefs, and shoals became part of the contested geopolitical imaginaries of space, ecology, and resources. The company's 3D presentations identified stretches of the river where rocky islets would be dynamited and dredging would be necessary, particularly along the navigational bottleneck between Chiang Saen to Chiang Khong where the Khon Phi Long reside. Clashes in conceptions of river space became apparent, as presentations by the company utilised technical engineering language, translated into Thai, that basically reduced the river to a 'shipping canal' in need of modification (Stone, 2019). This contrasts sharply with the notion of a riverscape that incorporates its multi local uses, sub-ecologies, and cultural connections to the river and its features.

Language and meaning are critical within environmental narratives, and so, when the Chinese company called the uncleared (non-channelised) sections of the river 'primitive', it provoked responses in favour of a free-flowing river from local activists. One Buddhist monk speaking at the Chiang Saen meeting is reported to have opposed the label of 'primitive', arguing "Thammachat (the natural system and state of the river) is not undeveloped", adding that to blow up rocks and dredge the river "without answering the concerns of local people would be the same as stealing" (Stone, 2019, n.p.). Similarly, the MRC CEO at the time, Dr. Pham (2017), wrote in the Bangkok Post that local activists and opponents to the navigational modification scheme should "stand back" and listen to "the waterway experts". "Waterway experts have clearly proven that in most of the case studies, 'clearing rapids' is barely a surgical intervention by which some rock outcrops and rocky riverbanks are chiselled (sometimes bedrock-deepened) and solitary rock outcrops in the navigation channel are removed and relocated." Rather, like 'the invisible' Xayaburi dam claims by senior Laos officials that the mainstream dam would be like a 'natural waterfall', a run-of-river dam with virtually no adverse impacts (Geheb et al., 2015), the surgical modifications to thousands of years of geomorphology for ship navigation purposes, would lead only to improved trade with virtually no adverse ecological or social impacts. Recognising that the terminology 'blasting the rapids' sounds 'extremely negative', MRC CEO Pham went on to suggest that navigation improvement was merely 'clearing rapids' for 'environmentally friendly' improvements in river trade. He concluded by saying: "we should all take a step back, reflect and view development in an objective and informed way." The contrast between the official institutional and corporate technocratic visions and counter-environmental narratives of activists reveal that position and place matter a great deal in the geopolitics that surround certain types of river modification and the locations of those ecological projects. There was virtually no opposition to the navigational improvements along the Myanmar–Laos section, but for over 20 years, there has been active community-based and non-governmental resistance to the navigational improvement scheme in Chiang Rai. In early 2020, the Thai Cabinet formally called for a cancellation of the Lancang-Mekong Navigation

Figure 10.2: River landscape diagram of physical elements in addition to rocks and reefs. Diagram adapted from Santasombat (2011). Source: Author.

Channel Improvement Project, representing a rare victory for community-based resistance in the Mekong Region (Lin and Grundy-Warr, 2020: 831).

In common with many other parts of the Mekong, appreciation of indigenous understandings of resources, local resource uses, seasonality, and vernacular readings of the landscape reveal that reducing values for very specific forms of transnational accumulation misses multiple ways of understanding the environment. Figure 10.2 reveals some common physical features apart from the rocks and reefs of the middle-lower part of the Mekong basin that play interrelated parts in the local riparian ecology and livelihoods. Yos Santasombat (2011), drawing upon studies by local community researchers, observed the significance of 'micro-ecosystems' for riparian communities. Villagers in Chiang Khong, for instance, divide their complex riverine ecosystem into 11 micro-ecosystems, that include 'pha' (rocks), 'khok' (deep pools near the riverbanks), 'don' (sandbars or islands), and 'hat' (sand or pebble beaches).

The removal of rocks and rapids would have ecological ramifications for all these micro-ecosystems. Some features are valuable habitats, spawning grounds, and refuges for many species of fish, aquatic plants and animals, and birds. Changes in flows and water levels due to the operations of hydropower dams affect micro-ecosystems throughout the river basin. Local knowledge is critical for governance at different places along the river system, and for understanding community rules and relations over resources and fisheries. Yong (2020) observed how traditional public fishing zones known as 'luang' were unmarked but known by local fishers, and based on ecological and seasonal spatial considerations. These relational resource geographies require intimate knowledge of socio-ecological practices and relations in seasonally changing river spaces. They are part of the multiple place-based commons that are connected by common environmental flows and seasonal rhythms, yet they are each unique in terms of in situ cultural–social-ecological relations.

Geopolitical channels and mitigation

> We are concerned that this project could proceed at the expense of biodiversity and the livelihoods of the Lao people who depend on fish that migrate up the Mekong River from Cambodia and Vietnam each year. We are especially concerned that the dam would block fish from migrating between Vietnam, Cambodia, Laos and Thailand, and that ultimately it would have a hugely negative impact on fisheries-based livelihoods in all four countries. (AMRC 2007).

> If built, the Don Sahong Dam will entirely block the Hou Sahong Channel, endangering fish migration throughout the Mekong region, jeopardizing food and livelihood security in Laos, Cambodia, Thailand and Vietnam. (International Rivers, 2014a).

> We are now very sure that with the mitigation measures we are going to do, [the dam] would have a very small impact to the downstream, or even to the upstream, about fish migration. As you can see, now we are working on the two channels to replace the Don Sahong channel for fish migrating. So, now there is good evidence, in a scientific way, that the fish can use other channels to migrate. That's why we don't think that it would have impact on downstream countries (Lao Official quoted in Mony, 2015).

These differing statements reveal that even the definition of a 'hou' or channel can become geopolitical by virtue of its border location, alleged values to biodiversity and fish migrations, and divergences between public officials, non-governmental watchdogs, and the concerned scientific community. In this case, the insistence that a dam on the Sahong channel was not a mainstream dam was in part related to the efforts of the dam proponents and the Lao Ministry of Energy and Mines to avoid a prolonged MRC Prior Consultation process, and proceed unilaterally with the construction of the dam. The Procedures for Notification, Prior Consultation and Agreement (PNPCA) are a process for MRC member states and relevant stakeholders to discuss and review benefits and risks of water-use projects on the mainstream, which may have significant cross-border impacts (MRC, 2020c).

Physical geography can thus be described in ways that fit a certain set of political economic objectives. A significant channel on the mainstream becomes 'one

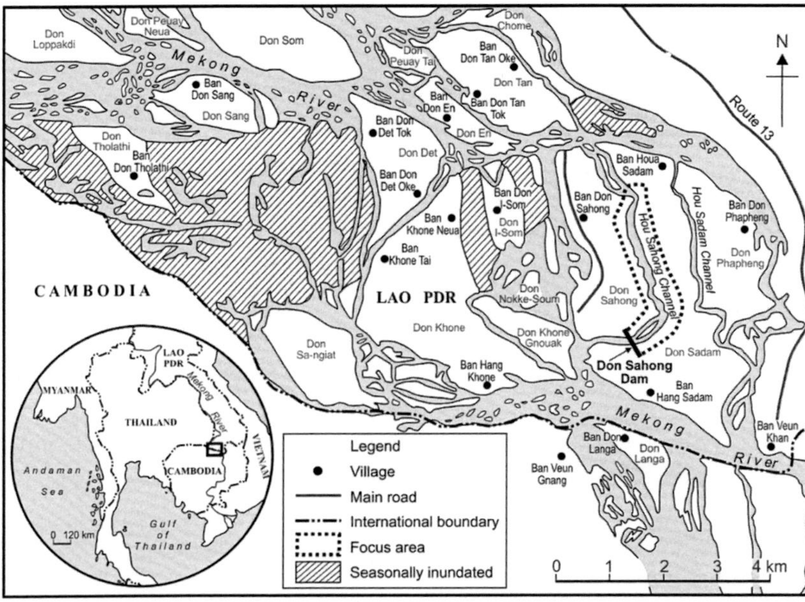

Figure 10.3: Map showing the location of the blocked Hou Sahong near the Lao-Cambodia border. Source: Author.

of many' and only a small fraction of the size of the mainstream to proponents, but 'one of the major' mainstream channels for migratory fish with substantial transboundary and downstream implications for opponents. For a while, this was a test case for the MRC's resolve in insisting upon the PNPCA process, although the organisation seemed to be sitting "on the proverbial transboundary fence" in terms of a clear geographical definition about what constitutes a mainstream project and a channel that is part of the mainstream (Grundy-Warr and Lin, 2020: 258). Geopolitics relates to the choice of dam site near the Cambodian boundary, with an EIA that does not focus enough on transboundary impacts.

On 28 July 2014, the Lao government decided to proceed with the PNPCA rather than go through a longer (and costlier) process of determining whether or not the Don Sahong constituted a 'mainstream dam'. For its part, "the MRC issued Development Partner statements of concern ... in order to promote a more cautious path of hydropower development on the Mekong mainstream" (Boer et al., 2016: 112). The 'hou' was blocked (Fig. 10.3), and the Don Sahong was built, opening a complex phase of uncertain outcomes and speculation about transboundary impacts. The dam's development, however, raises questions about environmental assessments and mitigation measures.

Grounded biophysical geopolitics is not just about the physical properties and ecological functions of physical features and ecological processes, but also about how natural elements and processes are "interpreted, manipulated, become part of developmental discourse or environmental claims and counter-claims in relation to specific ecological projects" (Grundy-Warr and Lin, 2020: 255). Thus, a geographical feature that is to be modified by a project is turned into a substantive geopolitical matter involving state officials, agencies, corporations, fishery experts, engineers, NGOs, and coalitions of concerned scientists. Political power resides with the dam backers as there was virtually no deliberative political space for arguing against the dam in Laos, and very limited space for oppositional voices to be heard from the Cambodian side.

Fisheries scientists and environmental NGOs were in little doubt that the 'hou' was a highly significant passage for both up- and down-stream migrations of fish at different times of the year, and that nearby channels besides Hou Sahong were much more limited as migratory pathways (Baran and Ratner, 2007; Baird, 2011). Indeed, the uniqueness of Hou Sahong as a vital migratory link was stressed as it was the only channel providing a year-round migration route (Baran et al., 2005), particularly "the only channel passable for late-dry season migrating species such as some large cyprinids" (Ryan, 2014: 3).

Mitigation and geo-engineered physical alterations became part of the arguments offered by the dam developers, arguing that channels such as the Hou Sadam and Hou Xang Pheuak could be deepened and widened to facilitate fish migrations through them, and "improve their viability as fish migration pathways" (Phonekhampheng et al., 2013: 22–23). There were, however, serious doubts about these claims and about the scope of engineering mitigation that was needed with a blocked Hou Sahong, and the pros and cons of nearby channels as migratory passages. WWF (2014) argued that Laos was adopting "an experimental approach" that was a threat to regional fisheries. "The approach is to continue experimentation until monitoring suggests fish are passing through the channels – a dangerous and unproven approach. The mitigation measures proposed by Mega First have never successfully been used in the Mekong basin or any other tropical river, and far from guarantee any success" (WWF, 2014: 5/7). In their appraisal of the Hou Sadam as a nearby channel for fish passage mitigation, Cowx et al. (2014: C21) noted: "[t]he location of Hou Sadam is remote from the Don Sahong channel and fishes attracted to the outfall of the hydropower development are unlikely to find Hou Sadam especially under low flow conditions." They noted that "only a small proportion of fish would use this channel", noting that it is "very narrow (e.g., 35 m) compared to Hou Sahong (e.g., 150 m) and possibly shallow". They observed significant sediment loading meaning that keeping the channel open would "require continual maintenance".

The deficiencies in the original EIA data, including supporting documents submitted by the dam developer, and doubts about the efficacy of proposed 'choke point' modifications to improve the passageway for migratory fish, reveal weaknesses in both environmental assessments and proposed mitigation works. The fact that a vital all-year migratory channel was chosen to be blocked ahead of other projects, and that proposals for modifying nearby channels were inadequate even prior to the dam's construction, reveals the politics of decision making in being able to literally 'channel' information in ways that will not halt the planned development going ahead.

Fishery specialists and consultants funded by the Don Sahong Power Company have continued to monitor the fish movements using underwater video monitoring in the modified Hou Sadam. In discussing passage effectiveness and efficiency, they reported: "No passage rates have ever been reported for Sadam Channel prior to its modification, but there is evidence that fishing pressure during past dry season migrations was so intense that successful passage was rare" (Hawkins et al., 2018: 242). They reported that "although channel modification improved fish passage efficiency, we estimated that artisanal fishers caught 79% of migrating fish in Sadam Channel, so fishing pressure remains the greatest risk to successful fish passage", but there was no analysis about the blocking of the Hou Sahong and whether or not this may have led to increased pressure on a narrower, modified fishing channel.

The dam and subsequent channel mitigation efforts have altered the geographies of fish migrations and created new fishing regulations in the dam project area. The implementation of fish trap prohibitions on key channels by the Don Sahong Fisheries Management Committee were "based on new understandings of what was obstructing fish migrations, the traps but not the dam" (Baird, 2024: 380), which became an environmental narrative specifically defined by the project's boundaries. The same study observes the technical problem of the less-than "fish friendly horizontal-bulb-type turbines used by the dam" which have "at least a five-percent mortality rate for downstream migrating fish", which if applied to all dams, "the cumulative impacts would be devastating" (Baird, 2024: 388). Thus, follow-up research on the peculiar ecological and social impacts of specific projects provides vital details to the larger canvas of hydropower capital altering hydrosocial relationships, ecologies, and geographies.

The MRC's (2020a) 358-page-long Hydropower Mitigation Guidelines raises some of the earlier critiques about the Don Sahong, and it admits: "locating a dam in one of many parallel channels might cause less environmental impacts, provided that fish migration is still supported by other channels. However, in the case of Don Sahong the dam is located in the Hou Sahong channel which many scientists claim to be the most important pathway for migratory fish in this section" (MRC, 2020c: 290). Just as developmental narratives close off alternative development pathways, it seems that dam decisions may prevent potentially better ecological alternatives from going ahead. As the MRC (2020c: 290) notes: "the Thako project that was planned for Khone Phapeng, would have produced a comparable energy output but without creating a barrier or impoundment. Nevertheless, it was not given serious consideration and was discarded based on the assumption that it would have negative impacts on tourism" (on the Lao side). The option of developing Thako and keeping the Hou Sahong open for year-round fish migrations in both directions, a better biological and ecological option, was not fully evaluated (Cowx et al., 2014).

Once a project starts—regardless of dissenting studies—we enter long periods of geo-engineered experimentation, speculation about impacts, and socio-ecological uncertainty. Boer et al. (2016: 105) noted "relatively little direct connection" between the MRC's "support for improving scientific knowledge, on the one hand, and the expectations that decisions on individual projects would be conditioned by a requirement that such knowledge be applied to the environmental assessments produced for each project, on the other hand." The use of only certain types of technical knowledge or specific interpretations of science to justify projects has long-term transborder implications on ecosystems, biodiversity, and social life associated with aquatic and riparian resources.

The Mekong River Basin has become a turbine-driving water machine that is fragmented by numerous cascade projects, tributary, and mainstream dams, which require further geo-engineered mitigation measures. Biophysical geopolitics draws attention to locally specific hydrosocial relations that relate to the transformation of bits of the natural environment. Each bit forms part of a basin-wide puzzle that requires analysis of the way in which they connect with and are shaped by national and regional environmental narratives and amalgams of political economic, technical, and financial power. Once a channel is blocked and a dam is built, there are irrevocable, if uncertain, biophysical transformations and a range of cross border social and environmental impacts. Subnational transboundary scales and the politics of position and place should be integral to environmental impact assessments, and follow-up research focusing on costs and benefits of ecological projects. The next vignette explores the local extinction of Irrawaddy dolphins in relation to subnational transboundary scales, the politics of position in a shared riverscape, and culture–nature interrelations.

Local extinction of Irrawaddy dolphins

As Beasley (2007: xvi) observed, "[a]s a result of small population sizes, strict habitat preferences, apparent high site fidelity, slow maturation rate, long calving intervals and most importantly their close proximity to human activities in freshwater systems, Irrawaddy dolphins are highly susceptible to anthropogenic impacts". In 2004, the Irrawaddy dolphin in the Mekong River was classified as 'critically endangered' on the IUCN Red List (Smith and Beasley, 2004) and classified as one of the 58 threatened species under the Cambodian Government's sub-decree on 'Determination of Types of Fisheries and Endangered Fisheries Product' in 2009. In addition, it is listed by CITES (Convention on International Trade in Endangered Species of Wild Fauna and Flora) as an Appendix I species (Eam et al., 2020: 6). Eam et al. (2020) surveyed the 180 km stretch of the Mekong inhabited by Irrawaddy dolphins, from the Khone Falls area and Stung Treng province to Kratie province, a stretch designated as the Mekong Dolphin Protection and Management Zone under a 2012 Cambodian Government Sub-Decree. This stretch is rich in biodiversity as it includes sub-ecologies relating to rapids, waterfalls, river islands and islets, flooded forests, sandbars, and deep pools, which are rich habitats for aquatic plants and animals. They estimated that, in 2020, the Irrawaddy dolphin population comprised 89 individuals, including both marked and unmarked individuals, similar to previous estimates of 92 individuals in 2017 (Ean et al., 2020).

The dolphin population in the river border segment between Stung Treng province in Cambodia and Champasak province in Laos had dwindled from 17 individuals in 1993 (Baird and Mounsouphom, 1994) to eight dolphins in 2007. This number fell to seven in 2009, six in 2012, three in 2018 and one with ID #35 in 2021. The impact of the Covid-19 pandemic may also have led to increased fishing activity in the dolphin range areas as people returned to their villages from the cities. Fishing threats to river dolphins are gillnet entanglement, illegal fishing practices including electrofishing, and poisoned bait (Thomas et al., 2019). Whilst the precise impacts of dam construction on the local dolphin population are unknown, "the changes due to dam construction were predicted to be much greater than the effects of climate change" (Ean et al., 2020: 5). For many years, scientists have stressed the dangers of large-scale infrastructure on rivers to bigger fish species, such as the Mekong Giant Catfish and mammals (Vander Zanden et al., 2004). More research is required to understand how changes in flow regime patterns and rhythms impact dolphins and river fish behaviour.

Reasons for the decline of the Irrawaddy dolphin population in the transborder river zone of Laos and Cambodia include the use of gillnets, electro-fishing, illegal cross-border fishing, genetic isolation as a sub-population, and the disturbances caused by the construction and operation of the nearby Don Sahong Hydropower Plant, including noise, changes in river flow, and loss of prey. Thus, proximity to human activity can endanger dolphins. The last dolphin, #35 (Phay et al., 2022) got its tail entangled in a fishing net, which caused injuries and difficulty in swimming, and eventually led to its death and extirpation.

Deep pools play a key ecological role, and photo identification studies reveal that dolphins move between deep pools in the Mekong, and the transboundary Chheu Teal deep pool plays a vital part in the ecology of the area straddling the Cambodian border near to Don Khone on the Lao side (Viravong et al., 2005). Deep pools play a vital role as dry season refugia and nurseries for many types of fish, and dolphins often used to be sighted around the transborder deep pool areas. A ban on gillnet fishing was enacted on the Cambodian side of the border, but not on the Lao side, and so gillnet fishing continued. As with rocks and reefs, recognition of the vital role that certain bits of geomorphology play in the river's ecosystem is often acknowledged in scientific reports and in the technical studies commissioned by the MRC, but regulation by relevant state authorities of these areas of particular ecological significance and sensitivity is typically weak.

Phay et al. (2022: 5) argue that there are indirect knock-on effects from the blockage of the Hou Sahong channel by the Don Sahong dam:

> The Don Sahong river is the main channel of the Mekong river in Lao where the majority of Lao fishermen conduct fishing operations. This channel is completely blocked by dam and fishermen of Lao come to do fishing along the transboundary river, using all types of fishing gears and practices. This has put more pressure on the Mekong fisheries biodiversity, leading to the extinction of the Mekong dolphin trans-boundary isolated sub-population.

Thus, the fishnet injury to #35 may at least be partially attributable to changes in fishing practices across the river border, although gillnets are a global problem for cetaceans (Reeves et al., 2009).

Prior to the Don Sahong dam's construction, the WWF raised concerns about its impacts on the survival of the last remaining dolphins in the area, particularly as a deep pool is close to the dam site, and the blocked Hou Sahong would likely reduce prey availability. WWF called for the immediate ban on gillnet fishing on the Lao side (WWF, 2012). When the Don Sahong dam site was under construction, a new danger to the dolphins emerged. Noisy construction activity, water current disturbance, and night lights were viewed as destabilising the dolphin habitat (Ryan, 2014). Such impacts were ignored or downplayed in the project's EIA, with vague information on fish migration and proposed mitigation measures based on scenarios that had not been scientifically proven (International Rivers, 2014b). This reveals an EIA that was deficient in exploring a variety of ecological impacts in this transboundary area affecting biodiversity and nonhuman life. However, precisely what ends up being environmentally assessed "tends to be deferred to later in the decision-making process" (Boer et al., 2016: 123), and thus, these technical practices end up deciding what "humans and non-human life, conditions and phenomena come to count" (Boer et al., 2016: 136). Potential cross-border socio-economic and ecological impacts of the Don Sahong dam construction were hardly addressed in its EIA. According to Yeo et al. (2024),

> the EIA failed to consider transboundary impacts on downstream Cambodian communities. 'We were upset about the building of the dam,' said Naby, a stall owner at Anlungsvay

market, who worried back then about its potential threat to the dolphin population and ecotourism—a key source of income for the community. 'Even though we sent a letter to the Lao government to protest against the construction, they didn't do anything'.

Hydropower and other developments have damaged ecotourism-related livelihoods and businesses connected with dolphins in Champasak province in southern Laos, while in Stung Treng of northern Cambodia, eco-tourism has almost vanished.

Political Animals, non-human charisma, love, and lessons for downstream conservation

Anthropomorphising wild animals is often a part of conservation efforts. Lorimer (2007) discusses the significance of 'nonhuman charisma' as a relational ontology with ethical and epistemological significance for ecological politics. He identifies three dimensions of nonhuman charisma: ecological, aesthetic, and corporeal charisma. Ecological charisma relates to the nature and frequency of any human–nonhuman encounters across space and time. These intersections are controlled by factors such as an organism's seasonality, migration patterns, day–night ecology, and distribution on land, in the air, or under the sea. The concurrence of its ecological rhythms with those of humans constitute a species' 'ecological charisma' (Lorimer, 2007: 917), and this relates to another aspect of nonhuman charisma, 'aesthetic charisma', "the distinguishing properties of an organism's behaviour and appearance that trigger particular emotions in those humans it encounters" (Lorimer, 2007: 921). Thus, Irrawaddy dolphins are characterised as being 'cute', 'playful', 'friendly', 'childlike', 'smiling', and characterised with other amiable human traits (Yeo et al., 2024). Proximity with dolphins can garner this sort of charisma, and it has often been the case that certain animals become 'flagship species' partly due to their aesthetic charisma. The third aspect of nonhuman charisma is "corporeal charisma which relates to the emotions ... experienced by self-described scientists in their everyday practices and interactions with their target organisms" (Lorimer, 2007: 922). Writing at a time when there was still a healthy population of Irrawaddy dolphins in southern Laos, Baird and Mounsouphom (1994: 163) observed that most people were friendly toward the dolphins, believing them to have saved people from harm when boats overturned or to protect people from crocodiles. They also reveal how local stories tie dolphins to human life. During her PhD thesis research, Beasley (2007: xii) provided names to each of the dolphin individuals which she studied and "spent many hundreds of hours photographing" hoping 'sincerely' that 'future generations' would continue to survive. Affectionate terms and personal names bestow dolphins with humanlike traits as distinct from scientific code numbers.

Thus, the last dolphin was named 'Sne' ('Love') by Bong (Uncle) Vanna, a boatman who became active in local ecotourism and conservation efforts relating to the dolphins. "Frustrated, helpless and dejected" was how Vanna felt when he saw the lifeless body of Sne (Fig. 10.4) in Preah Rumkel commune that shares the river border with Laos:

> He had never expected to witness the demise of the dolphins in his lifetime (...) Vanna had a special bond with the dolphins. Whenever he was out on his boat, they would come up to him and even play with the rudder. His favourite? Sne, the last dolphin that passed on last year. Vanna often gave the dolphins names that mirrored their personalities. 'Sne could recognise my voice', he said fondly, whilst showing us videos of their interactions. In his eyes, they were no different from his own children. Thus, akin to paying respect to the passing of a loved one, he held a funeral in remembrance of Sne. Drawing both the young and old, the community gathered together for a solemn farewell (...) Vanna dreams of having an annual Dolphin Festival to share the story of Preah Rumkel's dolphins with other provinces and urge communities to pursue conservation efforts seriously (Yeo et al., 2024).

Economically, the local extinction of dolphins had immediate impacts upon the local economy which was based upon ecotourism associated with dolphin watching. "Usually, tourists would come for the dolphins," said Tina. "But now, there are no more dolphins" (Yeo et al., 2024). Even before the passing of Sne, there were anxieties about the fate of the ecotourism in the area due to dam construction works. "We started to observe that the dolphins started to move to other places once the noisy construction works at Don Sahong began. If visitors come here looking for dolphins and they cannot see them they will say we are liars! Visitors will not come here again."[3] Another interviewee attributed one dead dolphin to dam works. "I found the dead body of a juvenile dolphin last year before Khmer New Year, and I am convinced it was due to polluted water from the dam construction."[4] This extirpation has meant that many people who previously earned incomes from eco-tourism have now had to seek alternatives, including out-migration.[5]

This case raises two intertwined issues, a local extinction with the demise of place-based livelihoods, and the links between nature and culture. Hobson (2007: 251) argues "that animals can and should be considered political geographical subjects on two counts": as "subjects of, and subject to, political practices" and because animals "can be considered affective political subjects". We should consider dolphins and migratory fish as "part of the heterogeneous networks that

[3] Author's interview with Bong Seng, a member of Preah Rumkel Commune, August 2018.

[4] Author's interview with Bong Vanna, Anlong Svay, Preah Rumkel, interviewed on 8 August 2018.

[5] Author's fieldwork in July 2024.

Figure 10.4: Finding the body of 'Sne', the last Irrawaddy dolphin in the Lao-Cambodia river border area. Photograph provided with permission of Phou Vanna.

constitute political life". The transboundary river where the extirpation occurred is in a uniquely beautiful area, with nearby waterfalls, rapids, islands, flooded forest, and a blend of Khmer–Lao river cultures. Bong Vanna sees the value in continuing eco-tourism without the dolphin as a means to educate the younger generation and visitors about the importance of conservation and value of ecology to livelihoods up and down the river, and to use the local extinction event to spur improved sub-national transboundary cooperation to protect biodiversity and natural resources.[6]

Pulsing ecosystems and hydrosocial relations

The flood pulse is the principal driving force helping to regulate productivity and ecosystem interaction within river–floodplain systems (Junk et al., 1989). It refers to the varying hydrological regimes between the wet and dry seasons in tropical and sub-tropical river–lake–floodplain systems, which drive their productivity and interactions amongst their living and non-living resources. A flood pulse produces a 'pulsing ecosystem' characterised by a 'moving littoral' throughout the aquatic–terrestrial transition zone (Junk and Wantzen, 2004). This has significant implications

[6] Interview with Vanna, 10 July 2024.

for productivity, life cycles, habits, food webs, and mobilities of aquatic organisms, including migratory fish patterns throughout the year. Chua et al. (2022: 21) observe: "[t]he annual flood pulse is the key to the regulation of the health of the floodplains, from fisheries to sediment replenishment."

Whilst the flood pulse should be viewed as a biophysical 'driver' throughout large parts of the Mekong River Basin, it is particularly crucial to the Tonle Sap Lake, which, due to the pulsing ecosystem, has been likened to the 'heartbeat' of the basin as a whole (Poulsen, 2003). Campbell et al. (2006) see the Tonle Sap as playing a critical role in absorbing large volumes of flood water from the Mekong River in the rainy season, reducing flooding along the mainstream, and in the dry season, releasing large water volumes to the mainstream, irrigating crops, and reducing salt intrusion in the Mekong Delta. With nutrients, regular natural flows, enough water volume, and good quality of water, the heart and geo-body are made healthy and highly productive in biodiversity and freshwater fisheries terms. The Tonle Sap remains vitally important as one of the most productive fisheries in the world despite evidence of fisheries decline (Lamberts, 2013; Yoshida et al., 2020). Writing at the early stages of the hydropower rush in the Mekong, Adamson et al. (2009) observed the 'remarkable' consistency and duration of the single wet-season flow peak and its "regularity throughout the huge basin". They warned that even 'small shifts' in 'timing' and 'flows' "could have deleterious effects on the ecology of this large river" (Adamson et al., 2009: 74), threatening a range of ecological flows and services to the lower Mekong Basin.

The hydropower 'binge' throughout the Mekong basin has contributed to a significant weakening of the river's flood pulse (Hecht et al., 2019), and the blocking of migration routes necessary for fish breeding. Viewed through a biophysical geopolitics lens, science relating to the flood pulse has not influenced policy and development planning. Hydrological studies and modelling of the changing flood pulse do not necessarily produce "clear and measurable societal goals for natural resources and economic development" (Arias et al., 2019: n.p.). However, the problem with hydrology-led, ecosystem-led, and other biophysically led messaging, based often upon years of data and sound methods, has not cut through to policymakers. As noted above, the MRC Secretariat has hundreds of scientific and technical reports detailing the relevance of flow regimes, sediments, seasonal flooding, wild-capture fisheries productivity, flooded forests, and other ecological services, but in three decades of hydropower expansion, these reports have done little to influence policy directions.

Figure 10.5 shows formerly floating houses that are raised on stilts during the dry season. It is a way of life that is rapidly changing as the conditions on the Tonle Sap become insecure due to a less predictable onset and duration of the flood period, and a weakened flood pulse in some years, producing longer than

Figure 10.5: Chhnok Tru commune, Kampong Chhnang (Cambodia). These are houses that were floating on Tonle Sap lake in the wet season, but raised up to dry land fringing the lake in the dry season. For this Cham Muslim group, the seasonal pattern was like this until they were instructed to move their homes permanently inland away from the lake. Source: Author.

normal dry spells. The houses shown in Figure 10.5 no longer have this seasonal lifestyle, as they were instructed by commune authorities, under a provincial instruction, to move their homes inland away from the lake proper. This has generated new expenses to move their boat homes and put them on high pillars to avoid inundation in the wet season, and created challenges for local people whose livelihoods have been dependent on aquatic resources, fishing, and living on and by water. Whilst pressures on lake resources and fisheries are great, the solution is not simply to remove people from the lake. It has caused people to take out loans as moving inland is costly, and many water-dwellers do not own land and need to supplement their water-dependent lifestyles. Being moved a kilometre from the lake means having to travel to their boats and spend more money on fuel.[7] For life-long fishers, the move from the lake is a traumatic experience that produces

[7] Author's fieldwork in June and September 2018, February, June 2019 in Chhnok Tru, Kampong Chhnang Province, and in Kampong Phluk community and Kampong Khleang commune, Siem Reap Province, Cambodia, 20-30 September 2024.

a social detachment from their old way of life into one of precarity and potential indebtedness (Grundy-Warr and Lin, 2020).

This situation of people being moved from the lake by authorities, or voluntarily due to pressures on aquatic livelihoods, may be repeated many thousands of times as changes to the flow regime occur due largely to the cumulative impacts of proximate and upstream ecological projects, particularly hydropower. Anthropogenic changes more immediately felt than climate change produce a "cascade of effects of rhythm loss" (Wantzen, 2022). This results from losing ecosystem services and weakening the hydrosocial ties that are integral to river cultures, and leading to either coerced displacements or environmental migrations away from the Tonle Sap (Sithirith and Grundy-Warr, 2024).

The Tonle Sap is far from being a rural backwater. It is, in fact, a densely populated rural zone with peri-urban areas nearer to provincial centres. There are over three million people living in the floodplain areas around the lake, with a high density of farming–fishing villages. A total of 800,000 people live within 170 seasonally inundated stand-stilt villages, and floating villages on the lake proper.[8] The Tonle Sap is a lake shared by six provinces, and connects to Phnom Penh municipality via the Tonle Sap River. The pulsing ecosystem of the lake with its pronounced wet and dry seasons leads to significant transformations in the landscape as the water surface in the dry season is 250,000 to 300,000 hectares which expands with annual flooding to a water surface area of 1,000,000 to 1,600,000 hectares.[9] Even as the flow regime and annual flood pulse are less regular due to multiple megaprojects, it is still a regulator of a multitude of livelihood and socio-economic activities.

Floating towns such as Kampong Luang in Pursat province, and the land–water-based settlements of Kampong Phluk and Kampong Khleang in Siem Reap province, alongside hundreds of villages within the floodplain, are indicative of water-linked urbanisation that is still reliant upon fishing, wetland resources, and aquatic agriculture around the Tonle Sap. Kampong Khleang commune comprises ten closely packed villages that coalesce together with 2,869 families and nearly 11,000 people.[10] Yet the economy is water based with a majority of households dependent upon capture fisheries. The message to policymakers should be clear: the Tonle Sap is a 'heartbeat' of a biodiverse, economically productive and densely populated rural ecosystem, and without a massive restructuring of the economy to provide alternative livelihoods, it remains essential for future generations in relation to their economic and environmental security.

[8] Information from the Fisheries Action Coalition Team (FACT), 22 September 2024.

[9] Data from FACT.

[10] Figures from commune authority, Kampong Khleang, September 2024.

Repositioning biophysical geopolitics and hydrosocial relations: A politics of brute force or persuasion?

National policymakers, water governance intergovernmental mechanisms, the MRC, the LMC, development agencies, engineering consultants, and stakeholders in the hydropower industry and river trade navigation have sought to 'develop' water to drive turbines, increase trade, generate revenues, and make profits. At sub-regional level, the geopolitical and geo-economic forms of cross-border linkage sponsored through the Greater Mekong Sub-region and LMC frameworks have promoted many state–SOE–capital ventures, including hydropower. Two decades ago, the MRC (2003: 6) noted that "development activities in the river system almost always result in simplification, even obliteration, of ecosystem diversity". State agencies, companies, and financial and development institutions have facilitated and accelerated the appropriation of water for energy away from the multiple communities whose lived spaces and life-worlds are very dependent on the flood pulse, floodplains, and freshwater ecosystems.

At the ASEAN (Association of Southeast Asian Nations) level, the lure of hydropower as 'sustainable energy' has been aided by the expansion of the ASEAN Power Grid that is legitimised in the media as a way to 'combat climate change'. Malaysia and Singapore are now part of a power integration project with Laos and Thailand. Channel News Asia excitedly reported that: "The Lao PDR-Thailand-Malaysia-Singapore Power Integration Project ... will import up to 100 megawatts (MW) of renewable hydropower using existing interconnections, marking the first multilateral cross-border electricity trade involving four ASEAN countries" (Ng, 2022). Energy security narratives have long been central to China's FDI and resource diplomacy in Southeast Asia (Zhao, 2015) and since 2015, the Lancang-Mekong cooperation framework has also facilitated Chinese energy companies to expand their regional activities.

Regional hydropower is pushed by "powerful political-economic interests ... who benefit materially from hydropower development" (Fox and Sneddon, 2019: 16). As the same authors acknowledge, "the types of knowledge" that are "privileged in public knowledge controversies over hydropower development ... are the product of brute political economic interests". Nearly three decades ago, Tyson Roberts argued that the vested hydropower interests were engaging in 'fluvicide' literally killing the hydro-ecological system, arguing that "tropical rivers with hydropower dams are not renewable, natural, or sustainable resources" (Roberts, 1996: 1). The sorts of knowledge that stress ecosystem integrity, the values of freshwater wild-capture fisheries, and the spatial–temporal rhythms of the annual flood pulse do not square with the agendas or material and technocratic interests of private and state-owned power companies, national energy ministries and institutions, or state officials benefiting from hydropower's expansion. Brute political economic forces have overshadowed deliberative politics concerning hydrosocial positional and place-based politics at sub-basin transboundary and sub-national levels.

There are other geopolitical impediments impeding moves to more socially and environmentally just forms of water governance. Water governance in the Mekong has long been hampered by 'scalar disconnects' within the jurisdictional landscape with differing national and subnational institutions, often with competing and contradictory agendas, approaches, mandates, and priorities within and across Mekong borders (Suhardiman et al., 2012), and "institutional arrangements that are inevitably redundant, incomplete, and only partially coordinated across levels and scales" (Lebel et al., 2005, n.p.).

Scalar disconnects of a biophysical and cultural politics nature relate to each of the vignettes in this chapter. Efforts to turn the upper mid-Mekong into an international waterway for ships were viewed as a means to expand trade, particularly Chinese trade, down the Mekong. The dynamiting of rocks and reefs met with community-based resistance in Chiang Rai, up-scaled as a regional environmental politics issue with the support of the Southeast Asian Rivers Network and International Rivers, among others. This example helps illustrate that technocratic instrumental values of nature and regional geo-economic imaginaries can be challenged at subnational and even transnational scales. Cultures sit in both places and nature, and place-based identities can sometimes be linked to the myriad ways in which people give meaning to nature that are beyond purely monetary and commodity forms (Harvey, 1996). The discussion about the choice of the Don Sahong among other channels for the siting of a hydropower dam is geopolitical not only in terms of its international boundary location, but in terms of the upstream and downstream fish, sediment, and other mobilities that are altered too. This has implications for a range of social and ecological issues, especially traditional fisheries on both sides of the border (Baird, 2024). Environmental assessments and engineering mitigation are part of "the politics of the technical" (Boer et al., 2016), and often prescribed by the authorities and companies who stand to profit from them. Complex local social and physical geographies altered and impacted should be part of bigger-scale investigations into the supposed "sustainability" of dams and other water-using megaprojects.

The death of Sne signalled the local extinction of the Irrawaddy dolphin at a subnational transboundary scale in the river system. In this case, we see the tragedy of a local extinction, with adverse social consequences. Lessons from this story reveal values that are not just monetary as they relate to nonhuman charisma and complex animal–human relations, which can act as valuable lessons to help strengthen conservation efforts involving riparian communities elsewhere in the Mekong. Conservation efforts often rely on ideas of 'flagship' species (such as the Irrawaddy dolphin or Giant Catfish) but equally important are cultural values and economic ones linked to earnings that can be derived from helping to monitor and protect species, and from conservation-based eco-tourism.

The Tonle Sap is a vital organ of the Mekong hydro-body, and one that is vital for water-linked economies, livelihoods, social relations, and current and future ecological security. If the 'beating heart' of the annual flood pulse becomes irregular and dysfunctional, millions of livelihoods will be at risk, and food security based upon freshwater fisheries will be jeopardised. People in the larger terrestrial–aquatic zone will have few alternatives to make a living if water resources become degraded to a point where they can no longer support land-and-water and water-based communities. They would be forced to migrate to Phnom Penh or to Thailand and elsewhere, leaving the communities of the lake and their unique wetland cultures to decline. Whole ways of life would disappear with enormous social, economic, and environmental costs and creating new political challenges for the region.

It is imperative that planners and policymakers look seriously at the cumulative and particular social and environmental impacts of a reliance upon centralised, technocratic, and coercive forms of water development. The political economic outlook should encompass dangers to regional cohesion and to society posed by weakened and dysfunctional biophysical processes. Calls past and present for more integrated and deliberative modes of water governance (Lebel et al., 2005; Suhardiman and Geheb, 2021), and for "alternative ways of being with water, ways of being that take into account riparian environments as they are, not simply as metrics upon a spreadsheet" (Johnson, 2024: 408–409) should be heeded. Perhaps a more effective form of persuasion for political elites, rather than stressing the vital socio-ecological linkages relating to 'living with water', will come from highlighting the potential and actual social conflicts, societal risks, and lost resource economies of environmental insecurity resulting from 'un' sustainable development. Whatever the case, there are a multitude of socio-ecological challenges across the Mekong Basin—and connecting communities, non-state actors, academics, and concerned international actors regionally at subnational and transnational levels, however fraught with state-imposed geopolitical obstacles, remains vital for healthier water governance.

References

Adamson, P.T., Rutherford, I.D., Peel, M.C. and Conlan, I.A. 2009. The hydrology of the Mekong River. In Campbell, I.C. (ed.), *The Mekong: biophysical environment of an international river basin*, 53–76. Amsterdam: Academic Press.

Allen, J. 2016. *Topologies of power: beyond territory and networks*. London: Routledge.

AMRC (Australian Mekong Resource Centre). 2007. Scientists concerned for the sustainable development of the Mekong River. AMRC, University of Sydney, School of Geosciences, Australia, 25 May 2007.

Arias, M.E., Holtgrieve, G.W., Ngor, P.B., Dang, T.D. and Piman, T. 2019. Maintaining perspective of ongoing environmental change in the Mekong floodplains. *Current Opinion in Environmental Sustainability* 37(2019): 1–7, https://doi.org/10.1016/j.cosust.2019.01.002.

Baird, I.G. 2011. The Don Sahong Dam: potential impacts on regional fish migrations, livelihoods, and human health. *Critical Asian Studies* 43(2): 211–235, https://doi.org/10.1080/14672715.2011.570567.

Baird, I.G. 2024. The Don Sahong Dam in Laos: political ecology, infrastructure, and the changing spatialities of impacts on fish and people. *Pacific Affairs*, 97 (2): 365–390, https://doi/10.5509/2024972-art1.

Baird, I.G. and Hogan, Z.S. 2023. Hydropower dam development and fish biodiversity in the Mekong River Basin: a review. *Water* 15(7): 1352, https://doi.org/10.3390/w15071352.

Baird, I.G. and Mounsouphom, B. 1994. Irrawaddy dolphins (*Orcaella Brevirostris*) in southern Lao PDR and northeastern Cambodia. *Natural History Bulletin of Siam Society* 42(1994): 159–175, https://archive.org/details/nhbss_042_2h_baird_irrawaddydolphinsor.

Baran, E. and Ratner, B.D. 2007. The Don Sahong dam and Mekong fisheries. Science Brief. Penang: WorldFish Center, https://digitalarchive.worldfishcenter.org/bitstream/handle/20.500.12348/1670/DonSahong-final.pdf?sequence1= (accessed 12 March 2024).

Baran, E. and Myschowoda, C. 2009. Dams and fisheries in the Mekong Basin. Aquatic Ecosystem Health and Management 12 (3): 227–234, https://doi/10.1080/14634980903149902.

Baran E., Baird I.G. and Cans, G. 2005. Fisheries bioecology at the Khone Falls (Mekong River, Southern Laos), Phnom Penh: WorldFish Centre, https://digitalarchive.worldfishcenter.org/bitstream/handle/20.500.12348/1942/Baran%20Baird%20Cans%202005%20Khone%20Falls%20fisheries.pdf?sequence1= (accessed 6 January 2024).

Basist, A. and Williams, C. 2020. Measuring the quantity of water flowing through the upper Mekong Basin under natural (unimpeded) conditions. Bangkok: Sustainable Infrastructure Partnership, https://558353b6-da87-4596-a181-b1f20782dd18.filesusr.com/ugd/bae95b_0e0f87104d-c8482b99ec91601d853122.pdf?index=true (accessed 15 January 2024).

Beasley, I.L. 2007. Conservation of the Irrawaddy Dolphin, Orcaella Brevirostris (Owen in Gray, 1866) in the Mekong River: Biological and Social Considerations Influencing Management, PhD dissertation, School of Earth and Environmental Sciences, James Cook University, Townsville, Australia, https://researchonline.jcu.edu.au/2038/10/JCU_2038_Beasley_2007_thesis.pdf (accessed 2 April 2025).

Bennett, J. 2010. *Vibrant matter: a political ecology of things*. Durham (NC): Duke University Press.

Biba, S. 2012. China's continuous dam-building on the Mekong River. *Journal of Contemporary Asia* 42(4): 603–628, https://doi.org/10.1080/00472336.2012.712257.

Biggs, D. 2010. *Quagmire: nation-building and nature in the Mekong Delta*. Seattle (WA): University of Washington Press.

Boer, B., Hirsh, P., Johns, F., Saul B. and Scurrah, N. 2016. *The Mekong: a socio-legal approach to river basin development*. London: Earthscan.

Campbell, I.C. (ed.). 2009. *The Mekong biophysical environment of an international river basin*. New York (NY): Academic Press.

Campbell, I., Poole, C., Giesen, W. and Valbo-Jørgensen, J. 2006. Species diversity and ecology of Tonle Sap Great Lake, Cambodia. *Aquatic Sciences* 68 (3): 355–373, https://doi.org/10.1007/s00027-006-0855-0.

Campbell, I. and Barlow, C. 2020. Hydropower development and the loss of fisheries in the Mekong River Basin. *Frontiers in Environmental Science* 8(2020): 566509, https://doi.org/10.3389/fenvs.2020.566509.

Castree, N. 2003. The geopolitics of nature. In Agnew, J., Mitchell, K. and Toal, G. (eds.), *A Companion to Political Geography*, 421–439. Oxford: Basil Blackwell.

Castree N. and Braun B. (eds.). 2001. *Social nature. theory, practice, politics*. Oxford: Basil Blackwell.

Chann, S., Mahanty, S. and Chamberlin, K. 2024. Squeezed between land and water: rupture, frontier-making, and resource conflicts at Cambodia's Lower Sesan 2 hydropower dam. *Pacific Affairs*, 97 (2): 313–342, https://doi.org/10.5509/2024972-art5.

Chua, S.D.X, Lu, X.X, Oeurng, C, Sok, T, and Grundy-Warr, C. 2022. Drastic decline of flood pulse in the Cambodian floodplains (Mekong River and Tonle Sap system). *Hydrology and Earth System Sciences* 26 (3): 609–625, https://doi.org/10.5194/hess-26-609-2022.

Clark, N. 2013. Geoengineering and geologic politics *Environment and Planning A: Economy and Space* 45(12), 2825–2832, https://doi.org/10.1068/a45646.

Cowx I., Mallen-Cooper M., Schmutz S., Bunthang T., Praxaysambath B., Sricharoendham B. and Vi An V. 2014. Prior Consultation for the Proposed Don Sahong Hydropower Project. Annex C: Technical Assessment Report. Fish passage and fisheries ecology. Based on submitted EIA Project Documents.

Dore, J. and Lebel, L. 2010. Gaining public acceptance: a critical strategic priority of the World Commission on Dams. *Water Alternatives* 3(2): 124–141, https://www.water-alternatives.org/index.php/alldoc/articles/vol3/v3issue2/86-a3-2-9/file.

Eam, S., Phay, S., Hang, S., Tan, S., Lor, K., Det, C. and Phorn, S. 2020. The monitoring of Irrawaddy dolphin population in the Mekong River: the long-term population monitoring based on mark-resight models. FiA/WWF- Cambodia Technical Report, https://wwfasia.awsassets.panda.org/downloads/the_2020_mekong_dolphin_population_report_final.pdf (accessed 29 March 2026).

Escobar, A. 2001. Culture sits in places: reflections on globalism and subaltern strategies of localization. *Political Geography* 20(2): 139–174, https://doi.org/10.1016/S0962-6298(00)00064-0.

Forsyth, T. and Walker, A. 2008. *Forest guardians, forest destroyers: the politics of environmental knowledge in northern Thailand.* Seattle (WA): University of Washington Press.

Fox, C. and Sneddon, C. 2019. Political borders, epistemological boundaries, and contested knowledges: constructing dams and narratives in the Mekong River Basin. *Water* 11(3): 413, https://doi.org/10.3390/w11103041.

Friend, R.M. 2009. Fishing for influence: fisheries science and evidence in water resources development in the Mekong Basin. *Water Alternatives* 2 (2): 167–182, https://www.water-alternatives.org/index.php/alldoc/articles/vol2/v2issue2/49-a2-2-1/file.

Geheb, K. and Suhardiman, D. 2019. The political ecology of hydropower in the Mekong River Basin. *Current Opinion in Environmental Sustainability* 37: 8–13, https://doi.org/10.1016/j.cosust.2019.02.001.

Geheb, K., West, N. and Matthews, N. 2015. The invisible dam: hydropower and its narration in the Lao Peoples' Democratic Republic. In Matthews, N. and Geheb, K. (eds.), *Hydropower development in the Mekong Region: political, socio-economic and environmental perspectives*, 101–126. London: Earthscan.

Gong, X. 2022 Words can speak louder than actions: examining China's discourse approach in Mekong governance. *Asia Policy* 17(2): 21–27, https://www.rsis.edu.sg/wp-content/uploads/2022/05/Gong-Xue.pdf.

Glassman, J. 2010. *Bounding the Mekong: the Asian Development Bank, China and Thailand.* Honolulu (HI): University of Hawai'i Press.

Grumbine, R.E. 2010. *Where the dragon meets the angry river: nature and power in the People's Republic of China.* Washington (DC): Island Press.

Grumbine, R.E., Dore, J. and Xu, J. 2012. Mekong hydropower: drivers of change and governance challenges. *Frontiers in Ecology and the Environment* 10(2): 91–98, https://doi.org/10.1890/110146.

Grundy-Warr, C. 2017. B/ordering nature and biophysical geopolitics: A response to Hirsch. *Political Geography* 58(2017): 131–135, https://doi.org/10.1016/j.polgeo.2016.10.006.

Grundy-Warr, C. and Middleton, C. 2025. Dams, flows, and data: topologies of power and volumetric hydropolitics in the Mekong basin. In Dressler, W and Mostafanezhad, M. (eds.), Violent atmospheres: livelihoods and landscapes in crisis in Southeast Asia, 209–237. Honolulu (HI): University of Hawai'i Press.

Grundy-Warr, C. and Lin, S. 2020. The unseen transboundary commons that matter for Cambodia's inland fisheries: changing sediment flows in the Mekong hydrological flood pulse. *Asia Pacific Viewpoint* 61(2): 249–265, https://doi.org/10.1111/apv.12266.

Grundy-Warr, C., Sithirith, M. and Li, Y.M. 2015. Volumes, fluidity and flows: rethinking the 'nature' of political geography. *Political Geography* 45(2015): 93–95, https://doi.org/10.1016/j.polgeo.2014.03.002.

Grünwald, R. 2021. Lancang-Mekong Cooperation: overcoming the trust deficit on the Mekong. ISEAS Perspective 89. Singapore: ISEAS – Yusof Ishak Institute, https://www.iseas.edu.sg/wp-content/uploads/2021/06/ISEAS_Perspective_2021_89.pdf (accessed 29 March 2026).

Grünwald, R., Wang, W. and Feng, Y. 2022. politicization of the hydropower dams in the Lancang-Mekong Basin: a review of contemporary environmental challenges. Energies 15(5): 1682, https://doi.org/10.3390/en15051682.

Harvey, D. 1996. *Justice, nature and the geography of difference.* Oxford: Blackwell Publishing.

Hajer, M.A. 1995. *The politics of environmental discourse: ecological modernization and the policy process.* Oxford: Oxford University Press.

Hawkins, P.R., Hortle, K.G., Phommanivong, S. and Singsua, Y. 2018. Underwater video monitoring of fish passage in the Mekong River at Sadam Channel, Khone Falls, Laos. *River Research and Applications* 34(3): 232–243, https://doi.org/10.1002/rra.3239.

Hecht, J.S., Lacombe, G., Arias, M.E., Dang, T.D. and Piman, T. 2019. Hydropower dams of the Mekong River basin: A review of their hydrological impacts. *Journal of Hydrology* 568(2019): 285–300, https://doi.org/10.1016/j.jhydrol.2018.10.045.

Hensengerth, O. 2009. Transboundary river cooperation and the regional public good: the case of the Mekong River. *Contemporary Southeast Asia* 31 (2): 326–349, https://doi/10.1355/cs31-2f.

Hirsch, P. 2011. China and the cascading geopolitics of lower Mekong dams. *The Asia-Pacific Journal* 9(20): 1–5, https://apjjf.org/2011/9/20/philip-hirsch/3529/article.

Hirsch, P. 2016. The shifting regional geopolitics of Mekong dams. *Political Geography* 51(2016): 63–74, https://doi.org/10.1016/j.polgeo.2015.12.004.

Hoang, T.H. 2022. China's hydro politics through the Lancang-Mekong Cooperation. ISEAS Perspective Issue 116, 22 November 2022. Singapore: Institute of Southeast Asian Studies (Yusof Ishak Institute, https://www.iseas.edu.sg/wp-content/uploads/2022/10/ISEAS_Perspective_2022_116.pdf (accessed 13 May 2024).

Hobson, K. 2007. Political animals? On animals as subjects in an enlarged political geography. *Political Geography* 26(2007): 250–267, https://doi.org/10.1016/j.polgeo.2006.10.010.

Hogan, Z., Campbell, T., Weisberg., P.J., Null, S.E. and Chandra, S. 2018. Mekong fish need more than designed flows', an online response to Sabo et al., 2018, https://www.researchgate.net/publication/324721259_Mekong_fish_need_more_than_designed_flow (accessed 17 November 2023).

International Rivers 2014a. Don Sahong Dam threatens Mekong fisheries, 16 December 2014, https://riverresourcehub.org/resources/259-4/ (accessed 4 May 2024).

International Rivers 2014b. Independent Review of Don Sahong EIA, https://archive.internationalrivers.org/resources/independent-review-of-don-sahong-dam-eia-8241 (February 2014) (accessed 12 October 2023.)

Jasanoff, S. 2012. *Science and public reason.* New York (NY): Routledge.

Johnson, A.A. 2024. Hidden flows: hydropower and the rhythms of development on the Mekong. *Pacific Affairs* 97(2): 391–409, https://doi/10.5509/2024972-art2.

Junk, W.J, Bayley, P.B. and Sparks, R.E. 1989. The flood pulse concept in river-floodplain systems. *Canadian Special Publication of Fisheries and Aquatic Sciences* 106 (1): 110–127, https://www.researchgate.net/publication/256981220_The_Flood_Pulse_Concept_in_River-Floodplain_Systems (accessed 29 March 2026).

Junk, W.J. and Wantzen, K.M. 2004. The flood pulse concept: new aspects, approaches, and applications – an update. In Welcomme, R and Petr, T. (eds.), Proceedings of the Second International Symposium on the Management of Large Rivers for Fisheries, Vol. 2, 117–149. Vol.2, Food and Agriculture Organization (FAO) and Mekong River Commission (MRC). Bangkok: FAO Regional Office for Asia and the Pacific. RAP Publication 2004/16, https://openknowledge.fao.org/server/api/core/bitstreams/4a514ff7-c110-4a2c-b1ab-29e53ed8d67d/content (accessed 29 March 2026).

Kallio, M. and Fallon, A. 2020. Are China's dams on the Mekong causing downstream drought? The importance of scientific debate. *Critical Nature* 28 April 2020, https://www.csds-chula.org/publications/tag/Marko+Kallio (accessed 29 March 2026).

Ketelsen, T., Räsänen, T. and Sawdon, J. 2020a. Did China turn off the Lower Mekong? Why data matters for cooperation. Southeast Asia Globe 13 May 2020, https://southeastasiaglobe.com/china-mekong-river-flow/ (accessed 1 November 2021).

Ketelson, T., Sawdon, J. and Räsänen, T. 2020b. Monitoring the quantity of water flowing through the upper Mekong Basin under natural (unimpeded) conditions: rapid review. Australia-Mekong Partnership for Environmental Resources & Energy systems (AMPERES), https://onedrive.live.com/?authkey=%21AGOd8Qb3xSyPaFw&id=692043E535D0F3B2%21572&cid=692043E535D0F3B2&parId=root&parQt=sharedby&o=OneUp (accessed 28 March 2026).

Lamberts D. 2013. The role and significance of the flood pulse in the functioning and management of the Tonle Sap Ecosystem, Cambodia. PhD dissertation, KU Leuven, Belgium, https://lirias.kuleuven.be/retrieve/227130

Latour, B. 2004, *Politics of nature: how to bring sciences into democracy*. Translation by C. Porter. Cambridge: Harvard University Press.

Latour, B. 2005. *Reassembling the social: an introduction to Actor-Network Theory*. Oxford: Oxford University Press.

Law, J. 2009. Actor-Network Theory and material semiotics. In Turner, B.S. (ed.), *The new Blackwell companion to social theory*, 141–158. London: Blackwell.

Lebel, L., Garden, P. and Imamura, M. 2005. The politics of scale, position, and place in the governance of water resources in the Mekong Region. *Ecology and Society* 10(2): 18, https://www.ecologyandsociety.org/vol10/iss2/art18/.

Lin, S. and Grundy-Warr 2020. Navigating Sino-Thai 'rocky' bilateral ties: the geopolitics of riverine trade in the Greater Mekong Subregion. *Environment and Planning C: Politics and Space* 38(5): 826–833, https://doi.org/10.1177/2399654420911410e.

Linton, J. and Budds, J. 2014. The hydrosocial cycle: defining and mobilizing a relational-dialectical approach to water. *Geoforum* 57(2014): 170–180, https://doi.org/10.1016/j.geoforum.2013.10.008.

Lorimer, J. 2007. Nonhuman charisma. *Environment and Planning D: Society and Space* 25(5): 911–932, https://doi.org/10.1068/d71j.

Magee D. 2006. Powershed politics: Yunnan hydropower under Great Western Development. *The China Quarterly* 185(2006): 23–41, https://doi:10.1017/S0305741006000038.

McDonald, K., Bosshard, P. and Brewer, N. 2009. Exporting dams: China's hydropower industry goes global. *Journal of Environmental Management* 90(2009): S294–S302, https:// doi:10.1016/j.jenvman.2008.07.023.

Middleton, C. and Allouche, J. 2016. Watershed or powershed? Critical hydropolitics, China and the 'Lancang-Mekong Cooperation Framework.' *The International Spectator* 51(3): 100–117, https://doi.org/10.1080/03932729.2016.1209385.

Mony, S. 2015. Lao Official Defends Don Sahong Dam. *VOA Khmer* 19 January 2015, https://www.voacambodia.com/a/lao-official-defends-don-dahong-dam/2604483.html (accessed 4 November 2024).

MRC (Mekong River Commission). 2003. Biodiversity and Fisheries in the Mekong Basin, Mekong Development Series No. 2, Phnom Penh: MRC Secretariat, https://archive.iwlearn.net/mrcmekong.org/download/free_download/Mekong_Development_No2.pdf (accessed 29 March 2026).

MRC (Mekong River Commission). 2015. Fisheries habitat and yield in the lower Mekong. *MRC Technical Paper* 47. Vientiane: MRC Secretariat, http://cantholib.org.vn:81/Files/1126.PDF (accessed 29 March 2026).

MRC (Mekong River Commission). 2020a. The MRC hydropower mitigation guidelines: guidelines for hydropower environmental impact mitigation and risk management in the lower Mekong mainstream and tributaries (Volume 3), Vientiane: MRC Secretariat, https://doi.org/10.52107/mrc.ajg006.

MRC (Mekong River Commission). 2020b. China to provide the Mekong River Commission with year-round water data, https://www.mrcmekong.org/news-and-events/news/china-to-provide-the-mekong-river-commission-with-year-round-water-data/ (accessed 16 December 2023).

MRC (Mekong River Commission). 2020c. Understanding the 1995 Mekong Agreement and the Five MRC Procedures: a handbook. Vientiane, MRC, https://www.mrcmekong.org/wp-content/uploads/2020/05/Understanding-the1995-Mekong-Agreement-and-the-Five-MRC-Procedures-A-Handbook.pdf (accessed 4 November 2024).

MRC (Mekong River Commission). 2021. The integrated water resources management–based Basin Development Strategy for the Lower Mekong Basin 2021–2030 and the MRC Strategic Plan 2021–2025. Vientiane: MRC Secretariat, http://www.mrcmekong.org/publications/basin-development-strategy-2021-2030-and-mrc-strategic-plan-2021-2015/ (accessed 29 March 2026).

Ng, H.S. Singapore begins importing regional energy from Laos through Thailand and Malaysia. *Channel News Asia* (CNA), 23 June 2022, https://www.channelnewsasia.com/singapore/singapore-import-hydropower-renewable-energy-laos-through-thailand-malaysia-2766251 (accessed 8 October 2024).

Osborne, M. 2019. The fight to preserve the Khon Pi Luang rapids. Lowy Institute, https://www.lowyinstitute.org/the-interpreter/fight-preserve-khon-pi-luang-rapids (accessed 18 September 2024).

Ó Tuathail, G. and Agnew, J. 1992. Geopolitics and discourse. *Political Geography* 11(2): 190–204, https://doi.org/10.1016/0962-6298(92)90048-X.

O'Lear, S. 2018. *Environmental geopolitics*. Lanham: Rowman & Littlefield.

Pham, T.P. 2017. The other side of the Mekong development debate. Bangkok Post, 6 November, 2017, https://www.bangkokpost.com/opinion/opinion/1355259/the-other-side-of-the-mekong-development-debate (accessed 18 September 2024).

Phay S., Eam, S.U., Hang, S., Tan, S.B., Lor, K. and Det, C. 2022. Summary report on the Status of the transboundary dolphins between Cambodia and Lao's PDR. Phnom Penh: The Fisheries Administration, World Wide Fund for Nature (WWF) and IUCN Cetacean Specialist Group, https://iucn-csg.org/wp-content/uploads/2022/03/Summary-Report-on-the-Status-of-the-Transboundary-Dolphins.pdf (accessed 29 March 2026).

Phonekhampheng, O., Busfield, A. and Hawkins, P. 2013. Report on Fisheries Study in Hou Sahong, Hou Sadam and Houxang Pheuak (2010-2012). Annex D to the 2013 EIA (accessed 4 June 2024).

Poulsen, A.F., Poeu, O., Viravong, S., Suntornratana, U. and Thung, N.T. 2002. Fish Migrations of the Lower Mekong River Basin: implications for development, planning and environmental management. *MRC Technical Paper* No. 8. Phnom Penh: Mekong River Commission, https://archive.iwlearn.net/mrcmekong.org/download/free_download/Technical_paper8.pdf (accessed 24 June 2026).

Pukinskis, I. and Geheb, K. 2012. The impacts of dams on the fisheries of the Mekong. *State of Knowledge* 1. Vientiane: CGIAR Challenge Program on Water and Food, https://cgspace.cgiar.org/rest/bitstreams/24817/retrieve.

Räsänen, T. A., Varis, O., Scherer, L. and Kummu, M. 2018. Greenhouse gas emissions of hydropower in the Mekong River Basin. *Environmental Research Letters* 13(3): 034030, https://doi/10.1088/1748-9326/aaa817.

Reeves, R., Brownell, R.L. Jr., Gulland, F., Smith, B.D., Turvey, S.T. and Wang Ding, 2009. Assessment of mortality of Irrawaddy dolphins in the Mekong River and recommendations for a population recovery plan. Gland: International Union for the Conservation of Nature, https://iucn-csg.org/wp-content/uploads/2010/03/Mekong_Dolphin_Mortality_report_from_international_experts.pdf (accessed 29 March 2026).

Rice, S., Tyner, J.A., Munro-Stasiuk, M., Kimroy, S. and Sirik, S. 2016. Making land to make life: island making in the South China Sea and the biopolitics of geophysical transformation. *The Geographical Journal* 182(4): 444–448, doi: 10.1111/geoj.12173.

Roberts, T.R. 1996. Fluvicide: an independent environmental assessment of the Nam Theun 2 hydropower project in Laos, with particular reference to aquatic biology and fishes. Bangkok: Smithsonian Tropical Research Institute, https://riverresourcehub.org/wp-content/uploads/2004/09/tysonfluvicide0904.pdf.

Rowedder, S. 2020. Railroading land-linked Laos: China's regional profits, Laos' domestic costs? *Eurasian Geography and Economics* 61(2): 152–161, https://doi.org/10.1080/15387216.2019.1704813.

Ryan, G.E. 2014. The Don Sahong Dam and the Mekong River Dolphin: an updated review of the potential impacts of the Don Sahong Hydropower Project on the Mekong River's critically endangered Irrawaddy Dolphins (Orcaella brevirostris), WWF, https://wwf.panda.org/wwf_news/?215472/The-Don-Sahong-dam-and-the-Mekong-dolphin (accessed 12 October 2023).

Sabo, J.L., Ruhi, A., Holtgrieve, G.W., Elliott, V., Arias, M.E., Ngor, P.B., Räsänen, T.A. and So Nam 2018. Designing river flows to improve food security futures in the lower Mekong Basin. *Science* 358 (eaao1053), https://doi/10.1126/science.aao1053.

Santasombat, Y. 2011. *The river of life: changing ecosystems of the Mekong Region*. Chiang Mai: Mekong Press.

Sarkkula, J., Keskinen, M., Koponen, J., Kummu, M., Richey, J.E. and Varis, O. 2009. Hydropower in the Mekong Region: what are the likely impacts upon fisheries? In Molle, F., Foran, T. and Käkönen, M. (eds.), *Contested waterscapes in the Mekong Region: hydropower, livelihoods and governance*, 227–249. London: Earthscan.

Schofield, C. 2013. What is at stake in the South China Sea? Geographical and geopolitical considerations. In Beckman, R., Townsend-Gault, I., Schofield, C., Devenport, T. and Bernard, L. (eds.), Beyond territorial disputes in the South China Sea: legal frameworks for the joint development of hydrocarbon resources, 11–46. Singapore: NUS Centre for International Law.

SEARIN (Southeast Asian Rivers Network). 2003. Mekong rapids under fire. Chiang Mai: Southeast Asian Rivers Network, http://www.livingriversiam.org/images/pub/mek_rapid_book.pdf (accessed 4 November 2024).

Sithirith, M. and Grundy-Warr, C. 2024. The social flood pulse and socio-ecological transformation of the Tonle Sap. *Singapore Journal of Tropical Geography* 46(1): 67–94, https://doi.org/10.1111/sjtg.12573.

Sluiter, L. 1992. *The Mekong currency*. Bangkok: Project for Ecological Recovery.

Smith, B.D. and Beasley, I. 2004. *Orcaella brevirostris* (Mekong River subpopulation). The IUCN Red List of Threatened Species 2004: e.T44555A10919444, http://dx.doi.org/10.2305/IUCN.UK.2004.RLTS.T44555A10919444.en

Sneddon, C. 2007. Nature's materiality and the circuitous paths of accumulation: dispossession of freshwater fisheries in Cambodia. *Antipode* 39(1): 167–193, https://doi.org/10.1111/j.1467-8330.2007.00511.x.

Sneddon, C. 2015. *Concrete revolution: large dams, Cold War geopolitics, and the US Bureau of Reclamation*. Chicago (IL): The University of Chicago Press.

Sneddon, C. and Fox, C. 2006. Rethinking transboundary waters: a critical hydropolitics of the Mekong basin. *Political Geography*, 25 (2): 181–202, https://doi.org/10.1016/j.polgeo.2005.11.002.

Sneddon, C. and Fox, C. 2012. Inland capture fisheries and large river systems: a political economy of Mekong fisheries. *Journal of Agrarian Change* 12(2–3): 279–299, https://doi.org/10.1111/j.1471-0366.2011.00350.x.

Sneddon, C. and Fox, C. 2019. Political borders, epistemological boundaries, and contested knowledges: constructing dams and narratives in the Mekong River Basin. *Water* 11(3): 413; https://doi:10.3390/w11030413.

Soukhaphon, A., Baird, I.G. and Hogan, Z.S. 2021. The impacts of hydropower dams in the Mekong River Basin: A review. *Water*, 13 (3): 265, https://doi.org/10.3390/w13030265.

Stone, A. 2019. Chinese firm fails to convince locals over Mekong blasting. Dialogue Earth, https://dialogue.earth/en/business/11040-chinese-company-consults-locals-over-mekong-blasting/ (accessed 18 September 2024).

Suhardiman, D. and Geheb, K. 2021. Participation and politics in transboundary hydropower development: The case of the Pak Beng dam in Laos. Environmental Policy and Governance 32(4), 320–330, https://doi.org/10.1002/eet.1974.

Suhardiman, D., Giordano, M. and Molle, F. 2012. Scalar disconnect: The logic of transboundary water governance in the Mekong. *Society and Natural Resources* 25 (6): 572–586, https://doi/abs/10.1080/08941920.2011.604398.

Swyngedouw, E. 1997. Neither global nor local: 'glocalization' and the politics of scale. In K. Cox (ed.), *Spaces of globalization: reasserting the power the local*, 137–166. New York (NY): The Guilford Press.

Swyngedouw, E. 2004. *Social power and the urbanization of water: flows of power*. Oxford: Oxford University Press.

Tadaki, M. 2017. Rethinking the role of critique in physical geography. *The Canadian Geographer* 61 (1): 73–83, https://doi.org/10.1111/cag.12299.

Techawongtham, W. 2017. With river sacrifice, regime yields to Beijing. *Bangkok Post*, 19 May 2017, https://www.bangkokpost.com/opinion/opinion/1252186/with-river-sacrific-regime-yields-to-beijing (accessed 1 October 2024).

Thomas, P.O., Gulland, F.M.D., Reeves, R.R., Kreb, D., Ding, W., Smith, B., Malik, M.I., Ryan, G.E. and Phay, S. 2019. Electrofishing as a potential threat to freshwater cetaceans. *Endangered Species Research* 39(2019): 207–220, https://doi.org/10.3354/esr00962.

United States Department of State, 1962. *International boundary study: Laos-Thailand boundary*, No.20, 18 September 1962, Washington (DC): Office of the Geographer. Bureau of Intelligence and Research.

Vander Zanden, J., Hogan, Z., Moyle, P.B., May, B. and Baird, I.G. 2004. The imperilled giants of the Mekong. *American Scientist* 92 (3): 228–237, https://doi/10.1511/2004.47.929.

Viravong, S., Phounsavath, S., Photitay, C., Solyda, P., Sokheng, C., Kolding, J., Valbo Jørgensen, J. and Phoutavong, K. 2005. Deep pool survey 2003-2004 – Final Report. Final Report. Phnom Penh: Mekong River Commission, https://www.academia.edu/967353/Deep_pool_survey_2003_2004 (accessed 23 September 2024).

Vu, T-M. and Mayer, M. 2018. Hydropower infrastructure and regional order making in the Sub-Mekong region. *Revista Brasilera de Política Internacionale* 61(1): e014: https://doi.org/10.1590/0034-7329201800114.

Wantzen, K.M. 2022. River culture: how socio-ecological linkages to the rhythm of the waters develop, how they are lost, and how they can be regained. Geographical Journal 190(2): e12476, http://doi.org/10.1111/geoj.12476.

Williams, J.M. 2020. The hydropower myth. *Environmental Science and Pollution Research* 27(2020): 12882–12888, http://doi.org/10.1007/s11356-019-04657-6.

Whitehead, M., Jones, R. and Jones, M. 2007. *The nature of the state: excavating the political ecologies of the modern state*. Oxford: Oxford University Press.

Whitington, J. 2018. *Anthropogenic rivers: the production of uncertainty in Lao hydropower*. Ithaca (NY): Cornell University Press.

Wong, T. 2010. Making the Mekong: nature, region, postcoloniality. PhD Dissertation. Columbus (OH): Department of Geography, Ohio State University, http://rave.ohiolink.edu/etdc/view?acc_num=osu1275482743 (accessed 12 May 2023).

World Bank and ADB (Asian Development Bank). 2006. Mekong Water Resources Assistance Strategy (MWRAS). WB/ADB Joint Working Paper on Future Directions for Water Resources in the Mekong River basin. Washington (DC): The World Bank and the Asian Development Bank, https://documents1.worldbank.org/curated/en/156061468122956051/pdf/367600EAPoPo8810ESWorepor-toonoMWRAS.pdf (accessed 21 March 2024).

WWF (World Wildlife Fund). 2012. Laos' last chance to save last 6 river dolphins. 27 September 2012, https://wwf.panda.org/wwf_news/?206300/Laos-last-chance-to-save-last-6-river-dolphins (accessed 12 December 2023).

WWF (World Wildlife Fund). 2014. Don Sahong Dam – Sustainable Solutions Exist. WWF Brief. Phnom Penh: World Wildlife Fund, Cambodia, http://d2ouvy59podg6k.cloudfront.net/downloads/alternative_to_don_sahong__thako_project.pdf (accessed 4 November 2024).

Xiao, L., Wang, J., Wang, B. and Jiang, H. 2023. China's hydropower resources and development. *Sustainability* 15(5): 3940, https://doi.org/10.3390/su15053940.

Yeo, H.Q.K, Koh, H.D., Ong, Y.T.N., Heang, S. and Baen, S. 2024. From Preah Rumkel, with Love. Mekong Despatches, Department of Geography, National University of Singapore. Permissions from interviewees granted. Based on a field investigation in September-October, 2023.

Yong, M.L. 2020. Reclaiming community spaces in the Mekong River transboundary commons: shifting territorialities in Chiang Khong, Thailand. *Asia Pacific Viewpoint* 61(2): 203–218, https://doi.org/10.1111/apv.12257.

Yong, M.L., Grundy-Warr, C. and Lin, S. 2024. A politics of mobility and b/ordering in a changing riverscape in Cambodia. *Pacific Affairs*, 97(2): 285–312, https://doi.org/10.5509/2024972-art6.

Yong, M.L. and Gillespie, J. 2022. Towards relational geometries of public participation and hydropower governance in the Lower Mekong River Basin. *Political Geography* 99(2022): 102773, https://doi.org/10.1016/j.polgeo.2022.102773.

Yong, M.L. and Grundy-Warr, C. 2012. Tangled nets of discourse and turbines of development: Lower Mekong mainstream dam debates. *Third World Quarterly* 33(6): 1037–1058, https://doi.org/10.1080/01436597.2012.681501.

Yoshida, Y., Lee, H.S., Trung, B.H., Tran, H.-D., Lall, M.K., Kakar, K. and Xuan, T.D. 2020. Impacts of mainstream hydropower dams on fisheries and agriculture in lower Mekong Basin. *Sustainability* 12(6): 2408, https://doi.org/10.3390/su12062408.

Yu, Y., Bo, Y., Castelletti, A., Dumas, P., Gao, J., Cai, X., Liu, J., Kahil, T., Wada, Y., Hu, S., Liu, B., Zhou, F. and Zhao, J. 2024. Transboundary cooperation in infrastructure operation generates economic and environmental co-benefits in the Lancang-Mekong River Basin. *Nature Water* 2(6): 589–601, https://doi.org/10.1038/s44221-024-00246-1.

Zhao, H. 2015. *China and ASEAN: energy security, cooperation and competition*. Singapore: ISEAS and Yusof Ishak Institute.

Zhou, Y., Miao, Z. and Urban, F. 2020. China's leadership in the hydropower sector: identifying green windows of opportunity for technological catch-up. *Industrial and Corporate Change* 29 (5): 1319–1343, https://doi: 10.1093/icc/dtaa039.

Ziv, G., Baran, E., Nam So., Rodríguez-Iturbe, I. and Levin, S.A. 2012. Trading-off fish biodiversity, food security, and hydropower in the Mekong River Basin. *Proceedings of the National Academy of Sciences* 109 (15): 5609–5614, www.pnas.org/cgi/doi/10.1073/pnas.1201423109.

CHAPTER 11

The political geographies of Mekong River Basin hydropower development: A reflection

Ian G. Baird[1]

Abstract

Hydropower development on the mainstream Mekong River and its tributaries has been occurring rapidly for decades, involving an array of political geographies, operating through a range of different socially and politically produced scales. Here, I consider the important political geographies associated with hydropower development in the Mekong River Basin. I also examine regional and international political geographies, including those associated with dam project development and water, sediment, fish, energy, and financial flows. By using political geography as a lens for considering hydropower dam development in the Mekong River Basin, and considering scale and power relations, I demonstrate how hydropolitical assemblages are produced and their important implications for the future of the Mekong River Basin, including its human and nonhuman dependents.

Keywords: Mekong, hydropower dams, political geography, space, flows

Introduction

Political geography is the study of relationships between politics and space, including scale and power. It is attentive to uneven political circumstances manifested and experienced across spaces, along with the ways in which politics is affected by various kinds of spatial relations (Agnew, 1997; Jessop et al., 2008), and how politics affects space (Leitner et al., 2008), including scales associated with hydropower development. These scalar changes typically disadvantage certain groups, such as poorer rural people who tend to be marginalised by decision-making processes that are typically dominated by other groups less reliant on nature-based livelihoods, especially political bodies located in urban centres. Constructing political geography frameworks has proved useful in various contexts (Kurtz, 2003), including for examining hydropower dam development in the Mekong River Basin (see Bakker, 1999; Hirsch, 2016; Middleton, 2022), the topic of this chapter.

[1] Department of Geography, University of Wisconsin-Madison, USA.

The Mekong River Basin (MRB) is a geographically and politically complex space. The river is long (~4,900 km) and its basin is large (some 795,000 km²). It also has a complex geological history that has affected biodiversity and ecological processes (Baird et al., 2003; Rainboth, 2012). While six nations share the Mekong Basin's territory, dozens of ethno-linguistic minority nationalities also share this landscape. Four of these countries (Cambodia, Laos, Thailand, and Vietnam) are members of the basin's management body, the Mekong River Commission (MRC), while China and Myanmar are not. China argues that international law provides individual sovereign states with the right to use the water within their borders. From a basin-wide-ecosystem-and-livelihoods perspective, however, this stance is problematic. Another international framework that addresses the basin is the Lancang-Mekong Cooperation (LMC) framework, which addresses all of the basin's countries, but is a unilateral Chinese creation, whereas the MRC is an interstate organisation. Crucially, the LMC is geared toward promoting trade and investment, rather than promoting effective joint governance of the Lancang/Mekong River and its tributaries, for which it has been criticised (Middleton and Allouche, 2016). Indeed, by prioritising development cooperation and investment, certain sensitive but critical issues have been underemphasised or partially sidelined in favour of China's economic and political goals.

In this paper, I will consider a series of complex 'flows' within and outside the basin that are closely tied to its political geography, that generate certain situations and responses. For example, water, along with its rich sediment load, flows from upstream to downstream. Another flow is the large number of fish, comprising many different species that seasonally migrate up and down the river, and in and out of its tributaries. So too, there are flows of people, who have long variously traversed parts of the Mekong River and its tributaries, whether as subsistence-level fishers or large-scale commercial transport operations. Over the last decade, the political geographies of these flows have become even more complex due to the development of many hydropower dams. These dams have, furthermore, created new flows of electricity and finance in and out of the basin, within nation states but also across national borders, and even to places as far from the basin as Malaysia and Singapore.

This chapter considers hydropower dam development in the MRB from the perspective of political geography. To do this, I explicitly consider some of the most important political geographies associated with the MRB's hydropower development, and how they are variously connected. Ultimately, I argue that applying a political geography lens when examining the MRB can help to illuminate the uneven power relations, injustices, and short-sighted decision making associated with the development of hydropower in the basin. I also demonstrate how hydropolitical 'assemblages' are produced and their important implications for the future of the MRB, including its human and nonhuman dependents. While this chapter does not present new empirical information, its contribution lies in its synthesis of existing information, and the addition of more analysis, especially in relation to the hydropolitical geographies of various kinds of flows associated with hydropower dams in the basin.

This chapter draws on my own 30+ years of experience living, working, and conducting research on hydropower development in the basin. I also draw upon the words of many researchers, as well as local people who I have interacted with over the years. I begin by providing a brief history of hydropower development in the MRB. I then consider how hydropower development has altered various kinds of flows and movements within and outside of the basin. I then assess how new political geographies associated with the basin have led to new injustices in the region, while also exacerbating previously existing ones. Finally, I provide some concluding thoughts, and propose some ways that more just political geographies in the region can be promoted.

A brief history of hydropower development in the Mekong River Basin

The first large dams in the MRB were built in northeast Thailand, beginning with the Nam Pung in Sakon Nakhon Province in November 1965 (Molle et al., 2009). During this period, the Communist Party of Thailand was gaining influence in northeastern Thailand, and the United States wanted to promote modernisation and development as a way of countering the spread of communism.

The first large hydropower project developed in Laos was the Nam Ngum dam (now Nam Ngum 1), the construction of which was supported by a number of governments led by Japan, and was completed in 1971 (Hirsch, 2010). This dam was also intended to hamper the spread of communism in the country. There were plans for many other dams, including on the river's mainstream, but Cold War conflict prevented many of the projects envisioned from being constructed (Sneddon, 2015).

Hydropower dam development was once again ramped up in the 1990s, as part of an effort to promote development during a post-Cold War era. The Pak Mun Dam, in the northeastern Thai province of Ubon Ratchathani, was amongst the first dams to be developed during this period, despite widespread local opposition at different times (Foran and Manorom, 2009; Missingham, 2003). During the 1990s, the government of Thailand also began developing other medium and large irrigation dams as part of the Khong-Chi-Mun Scheme (Sneddon, 2003).

In the mid-1990s, dam construction in Laos also experienced a resurgence as a part of efforts to promote development during a post-Cold War period, with the support of multilateral banks, such as the Asian Development Bank (ADB), and private foreign capital support. This included the development of the Theun-Hinboun and Houay Ho Dams in central and southern Laos, respectively, and a small number of other dams (Whitington, 2018; IRN, 1999). In Thailand, it was getting politically more difficult to build dams due to widespread opposition (Kripakorn, 2007; Missingham, 2003), and the Thai government became increasingly interested in importing energy from Laos, where open opposition to dams was not (and still is not) permitted (International Rivers, 1999).

Hydropower construction was also underway in Vietnam's Central Highlands. In 1990, as Vietnam was promoting increased neoliberal economic reforms, the 12 MW Dray Hinh 1 dam was completed, while the 720 MW Yali Falls dam was completed in 2001 (Wyatt and Baird, 2007). Other dams in the upper Sesan and upper Srepok rivers soon followed. Meanwhile, the first dam on the mainstream Lancang (Mekong) River in China—the 1,670 MW Manwan—was completed in 1993 (Fu et al., 2008).

During 1998-1999, however, the Asian financial crisis resulted in some planned hydropower projects being cancelled; others were temporarily postponed, or scaled down. In addition, the critical findings of the widespread negative social and environmental costs of dams reported on by the World Commission on Dams (WCD, 2000) also caused large multilateral banks and governments to stop or delay funding new large dams in the early 2000s. In China, however, the situation differed, and hydropower dam development continued on the Lancang River (Hirsch, 2011). In 2005, the World Bank approved loan guarantees for the controversial Nam Theun 2 Hydropower Project in central Laos. The project was completed in 2010, and was, at the time, the largest dam in the lower Mekong River Basin (Porter and Shivakumar, 2011), although it was developed at a great cost to the environment and local people (Shoemaker and Robichaud, 2018). The dam was constructed during a period of increased scepticism about hydropower dam development following the World Commission of Dams report.

The situation changed dramatically in the 2010s. Hydropower dam development increased dramatically, particularly in Laos, China, and to some degree in Cambodia, during a period when China was gaining increased economic and political influence in mainland Southeast Asia. In Laos, as of February 2019, there were 61 hydropower dams, with the installed capacity of 7,207 MW, and the capacity to generate about 37,366 kilowatt-hours per year (Xinhua, 2019). Most projects were developed on the basis of the build–operate–transfer (BOT) or build–own–operate transfer (BOOT) model, in which private companies invest in projects for which they have a concession, usually for a period of 25–30 years. Once the concession period is over, the dam is handed over to the Lao government (Wyatt, 2004). In addition, 36 hydropower dams were under various stages of construction in early 2019. They were expected to contribute 4,184 MW in installed capacity by the time that they were completed by the end of 2020, although COVID-19 pandemic conditions slowed the development of some projects. Furthermore, the government of Laos was planning to begin developing another 10,000 MW of electricity by 2020, and an additional 20,000 MW by 2030 (Xinhua 2019), although progress slowed. Considering current projections of rapidly increasing installed capacity, and Laos' energy exports to neighbouring countries, Laos is realising its decades-old goal to become 'the battery of Southeast Asia'.

In Cambodia, the controversial Lower Sesan 2 Dam started construction in 2014, and was completed in 2018 with support from China's Belt and Road Initiative (Baird, 2016). Studies have also been conducted regarding the possible development of the Sambor and Stung Treng dams on the Mekong mainstream, although these projects have been deferred for at least a decade (Green and Baird, 2020), and the Ministry of Environment recently announced that Cambodia would not allow any dams to be developed on the mainstream Mekong River in Cambodia (Strangio, 2023).

Changing flows and movements in and beyond the Mekong River Basin

There are various kinds of flows and movements that are important for understanding the political geographies of the MRB. They can be conceptualised as assemblages, which emerge due to a complex combination of different influences and factors. In other words, wholes are constructed through the coming together and interactions of various individual parts. Assemblage theory is attentive to how these different parts are altered through this coming-together process (Anderson, 2021). In this section, I discuss various types of flows and movements, including some that have long existed, and others that have only emerged due to hydropower development. My focus is on hydropolitical assemblages, or assemblages that are constituted through an array of influences that affect the political geographies associated with the ways in which the flows develop and operate, and are sometimes blocked through particular processes, including those related to infrastructure development, but also due to complex and multifaceted decision-making processes. These have various scalar impacts and implications, some of which are discussed here.

The most obvious flow relates to water, which typically relies primarily on gravity to guide it from surfaces to small streams, to larger streams and rivers, to the Mekong River, and eventually to the Mekong Delta in Vietnam. These flows have seasonal ebbs, flows, and rhythms, based primarily on the monsoon rains, but also on seasonal snow, including ice and glacial melt in the uplands of southwestern China. More recently, however, these flows are being increasingly altered, especially the flood pulse system, both due to individual dams and the cumulative impacts of multiple dams (Hecht et al., 2019; Soukhaphon et al., 2021). Indeed, China now has the ability to fundamentally alter downstream Mekong flows, with the potential to cause serious social and environmental impacts (Eyler and Weatherby, 2020) because government agencies in China now have the ability to decide how such flows will be altered downstream. Political decisions have been made to disrupt fish migrations, sediment flows, and free-flowing rivers and their ecological benefits in order to generate energy for apparent economic reasons.

Other political decision making around dam construction also reflects the kinds of dams that should be built, and how to operate them, both on a daily or seasonal basis. Decisions are also made regarding how to prioritise dam operation, so as to satisfy certain priorities, values, and interests at various scales. For example, when less energy was being generated by the Don Sahong dam than expected, the Lao government made the decision to alter water flows by diverting more water through the dam's power station, thus sacrificing other values, including fish migrations and aesthetics associated with the Khone Phapeng Waterfalls (just downstream of the dam). Moreover, the presence of the Don Sahong Hydropower Company has effectively created new scales of governance, since some budgets associated with compensation but also local development now need to be approved not only by different scales of government but by the company as well.

While some imagine the Mekong as being a watershed, others see it as a 'power-shed' (see Hennig et al., 2016; Magee, 2006; Middleton and Allouche, 2016). The scale of electricity production is also important, with some advocating for small-scale hydropower production, based on environmental and social grounds, while others see large scale hydro as more suitable based on economic and technical arguments (Luukkanen et al., 2012). For the Lower Sesan 2 dam owners, the portion of the river below the dam is referred to as a 'discharge channel' rather than a river, demonstrating how dam developers attempt to change people's ideas about, and relationships with, rivers, and prioritise certain river uses and values over others. In addition, urban dwellers and national governments tend to benefit more from hydropower dam development than poorer people, as capital cities are typically more insulated from dam impacts. In addition, people from the cities tend to have higher formal education levels, and thus frequently benefit more from jobs associated with hydropower dams than those who dwell in rural areas. There is also a tendency for the revenues from dams to end up being accumulated and circulated in urban areas, thus benefiting urban economies the most. The provision of electricity supply to places that did not previously have access to it does, however, create spin-off job opportunities. Still, dam operators can make political decisions to maximise profits from dam operations by releasing different amounts of water at different times of the day and year, depending on demand and price (Baird and Quastel, 2015), although Power Purchase Agreements do limit operating flexibility, to a lesser or greater extent depending on the dam.

In a similar vein, dam operators often make decisions regarding how much water to seasonally store in their reservoirs, decisions that are often presented as technical, but are actually quite political. Governments and dam owners often make political decisions about how much compensation to provide, and to whom, relative to the negative impacts of projects. Decisions are also made by governments and dam owners about the geographical extent of impact areas and the amount of compensation to provide, decisions that frequently do not fairly include all of those who have been negatively impacted (Green and Baird, 2016; Scudder, 2005).

The second type of flow, i.e., sediment, goes hand in hand with water, but has different physical or material characteristics, even if its prominence, or lack of it, has also been affected by political decision making. Dams are very effective at 'trapping' sediments, thus disrupting historical patterns, and causing negative impacts downstream. Already a large portion of the sediment that flows down the Mekong is trapped behind dams (Chuenchum et al., 2023). Political decision making by governments and investors, often located far from the basin itself, have tended to diminish the importance of sediment when it comes to resource governance and management. Human values are crucial for determining the types of decisions that are made at different scales.

The third type of flow is fish. Large numbers of Mekong fish populations belonging to hundreds of species undergo seasonal and sometimes complex migrations up and down and in and out of the Mekong and its tributaries, triggered by hydrological conditions, such as rising or falling waters or lunar cycles (Baird et al., 2003). Humans living along the river have found ways to monitor these movements, especially upstream fish migrations, including using newly available cell phone technology (Baird and Manorom, 2019). This is important, because the MRB supports more fish species than any other river in Asia (Baird and Hogan, 2023, Kottelat et al., 2012) and one of the largest freshwater fisheries in the world. However, political decisions, such as to not collect adequate baseline data before constructing hydropower dams, and to de-emphasise the importance of wild capture fishers, has tended to result in less attention being paid to this important natural resource (Bush and Hirsch, 2005; Käkönen and Hirsch, 2009). Moreover, even when data are collected, such as regarding the impacts of dams on certain fisheries (see Middleton, 2022), these are insufficiently used to affect hydropower development policy, because they do not support the agendas of economically and politically powerful dam developers and their allies.

Many other species also move against water flows. These include migrating birds who follow fish and sometimes Irrawaddy dolphins migrating up rivers and streams, and of course, humans, who travel up and down rivers for a range of reasons, including following migrating fish (Baird and Manorom, 2019). Decisions regarding the place of these species in narratives and planning are undoubtedly political, and have often remained depoliticised in order to ensure that the interests of dam builders and their supporters are protected. Scale is also crucial, since large dams often prioritise national and regional scales of development and power, rather than other people impacted by particular projects, or nonhuman actors.

It is important to consider the relatively new directions of electricity flows that have emerged due to hydropower dam development, and the related building of new transmission lines within nation states and between them to transfer electricity across spaces. This is the fourth flow. These flows are affected by particular political decisions, including those related to urban planning in faraway cities (Marks and Zhang, 2019), and can have important environmental and social impacts due to the decisions regarding how to manage particular dams (Baird and Quastel, 2015). Along with energy planning more generally, these decisions are typically framed as being technical rather than political, but this is far from the truth. Some non-government organisations, such as the Bangkok-based Mekong Energy and Ecology Network (MEE Net), have challenged some important EGAT decisions, such as how much of an electricity reserve that Thailand should maintain, since such decisions have important implications in relation to the development of new hydropower projects. Political decisions regarding how to meet energy needs, whether through developing more solar or wind power, or relying more on hydropower energy imports are also important decisions. In addition, electricity flows have often been produced and enhanced through the development of infrastructure, such as transmission lines, that are crucial for the flow of electricity, again in the interests of modernised development and capital accumulation.

The fifth flow is finance and capital, which flow through new channels and in larger quantities due to dam development. The ease of this flow is promoted and enhanced, especially across borders, something that multilateral banks such as the World Bank and the ADB have been long advocating. In the MRB's early dam-building era, finance associated with dams mainly passed through or from national governments and foreign aid, either as grants or as loans from multilateral banks (Goldman, 2005; Middleton et al., 2009). Since developing the BOT and BOOT methods to attract private capital in the 1990s (Wyatt, 2004), the vast majority of finance for large dam construction has come from private international sources, especially Thai and Chinese banks (International Rivers, 2008; Middleton et al., 2013; 2015b). Chinese investment in dams has come with lower environmental and social standards, and lower requirements for local consultations and consent (Barney and Souksakoun, 2021; South China Morning Post, 2011; Urban, 2013). BOT legislation in Laos and Cambodia—supported by the ADB and the World Bank in the 1990s—has also played an important role in facilitating increased foreign investment in MRB hydropower development (Käkönen and Nygren, 2023). Thus, the sources and flows of finance and capital have shifted and expanded, and can be expected to shift further.

These shifts also influence the movement or flow of people. For example, large numbers of people have had to relocate due to hydropower dam development, and privileging dam development over the rights of people and nature has facilitated the flow of resettled people to new spaces, often to their detriment. The flow of largely temporary labour and expertise into areas where dams are being built has also been facilitated, or unblocked, through political decisions being made in the interests of dam building and state territorialisation. Indeed, both the MRC and the LMC Framework, because they privilege nation states and the rights of nation states to prioritise the interests of some over others through their structures, are certainly part of increasingly dominant state territorialisation processes. Crucially, the interests of capitalists now frequently coincide with those of nation states, thus leading to dam development that is both friendly to capital accumulation while concurrently strengthening state power in the areas where hydropower dams are developed.

Politicians often have their own particular scalar priorities and interests that conflict with those of other groups, particularly rural peasants and nonhuman actors, but also between central and local governments—a crucial aspect of hydropolitical assemblages. Sometimes, these interests are personal, and related to corruption and nepotism. They are also promoted in the name of anthropocentric modernised development and capital accumulation. This is particularly evident in Laos and Cambodia, where rural people and many nonhuman species are highly dependent on the Mekong and its tributaries for their survival, but tend to be marginalised by hydropower projects that negatively impact them. Indeed, wild-capture fisheries, despite playing a crucial role for people in the MRB, have long been marginalised at the political level (Friend et al., 2009; Orr et al., 2012), often because of an 'anti-politics' positioning. Here, the political aspects of hydropower decision making are denied, instead justifying hydropower dams as technically logical (Käkönen and Hirsch, 2009). This also greatly affects how fishery decline is politically framed (Bush and Hirsch, 2005). Is it caused by dams or overfishing? Such uncertainties are also frequently used to justify projects that cause serious negative impacts on fish and fisheries (Baird and Hogan, 2023; Whitington, 2018). Borrowing from Mahanty et al. (2023), some flows or movements, especially those related to water, sediment, and fish, have been critically 'ruptured', with their connectivity being severed in order to promote other flows, particularly those of energy and capital.

Crucially, some flows are being repressed through political decisions and legal and political frameworks, including ones that divide national and local governments. In particular, the flow of information about the negative social and environmental impacts of dams, including on livelihoods and the political decisions being made in support of hydropower development, are often repressed through various means, such as by limiting democratic rights and freedoms, including free speech and access to information; and also by controlling and influencing the media and other flows of information and exchange of ideas. Again, this is frequently done in support of the interests of small groups of powerful and politically connected individuals as well as the interests of capital accumulation.

Imagining more just political geographies in the Mekong Region

Water justice—and how it is achieved—is an important global issue (Sultana, 2018), and the Mekong is no exception. So far, however, hydropower development in the MRB has been far from just, for a range of reasons (Middleton et al., 2015a; Middleton and Pritchard, 2016), some of which have already been mentioned above. As stressed earlier, it is poorer rural people, who rely on the river most for their subsistence and livelihoods, who tend to benefit the least from dams, and are generally the most negatively impacted by them. At most, these people are expected to tolerate the total disruption of their lives, often for years or decades, without actually benefiting from dam development. Unsurprisingly, such communities tend to have fewer opportunities for their voices to be heard or to influence important decisions about whether to build new dams, and how to operate and compensate for those projects already in operation.

As should already be evident, the political geographies of hydropower development in the MRB, constituted through hydropolitical assemblages, are uneven and operate across a range of scales, from the international and regional to the national and finally to the local (Dore and Lebel, 2010; Hirsch and Wyatt, 2004; Sneddon, 2015), including those involving international bodies such as the Association of South East Asian Nations, the United Nations, and international multilateral banks, to the MRC and the LMC Framework (Hensengerth, 2015; Middleton and Allouche, 2016).

There are also other scales of political geographies, including at the national, provincial, district, village, neighbourhood, family, and individual levels, and variations of all of these. They are all important in different ways and at different times, and interact with each other in a multitude of ways. Institutions, alliances, and individuals always have uneven influences, and interact with hydropower dams and their impacts in various ways, and with a range of interests. Kin relationships have also long had an important influence on governance regimes in the Mekong Region, including decision making associated with natural resource management (Global Witness, 2016). Neopatrimonial political relations are common in the region, including in relation to hydropower dam development (Käkönen and Thuon, 2019), and are yet another important factor within hydropolitical assemblages, ones that affect decision-making processes at various scales.

Sometimes there are important contradictions between different scales of governance, even within individual countries. For example, local ethnic Lao people living on islands in the Mekong River in southern Laos have made wild-capture fisheries conservation and management a priority (Baird and Flaherty, 2005), but the Lao government has particularly promoted hydropower development that is destructive to fisheries (Soukhaphon et al., 2021).

Indigenous and local knowledge is also critically important for good and just natural resource management, and the degree to which certain types of knowledge are accepted or not by governing individuals and institutions is important. But these types of knowledge are often marginalised in various ways (Käkönen and Hirsch, 2009; Silvano et al., 2023). Mekong knowledge is also closely linked to different scales and political geographies (Grünwald, 2021; Middleton, 2022) affecting what is prioritised and what is not, and how different challenges are understood and approached.

Large hydropower dams have long been controversial, due to the dramatic ways that they alter rivers and impact dependent nature and human livelihoods (McCully, 2001; Scudder, 2005). It is therefore not surprising that resistance movements and networks of various types have emerged in relation to hydropower project development, including—most famously—Thailand's Pak Mun dam (Baird et al. 2020; Glassman, 2002; Missingham, 2003), but also various other projects in the region (Baird 2016; Hirsch, 2007).

The politics of scale is evident when it comes to how environmental assessments are conducted on hydropower dams in the MRB (Lebel, 2006). Indeed, the political geographies of environmental assessments, and how impact areas are defined and produced is critically important in the MRB (Green and Baird 2020; Matthews, 2013). Lamb (2014) has described the ways in which scale has been produced through environmental assessments in the Salween River Basin, something that is also true for the Mekong, albeit with differences in the details.

Undoubtedly, the transnational nature of the MRB and its commons are particularly challenging to address, especially since all the nations in the region are wary of giving up any sovereignty for the improved governance of the river and its basin (Geheb and Suhardiman, 2019; Middleton, 2022; Soukhaphon et al., 2021). This raises questions about the ways in which national sovereignty can be reconciled with good relational governance and local rights. The answers to such questions are not easy, given that the world is politically organised in ways that tend to prioritise national scales of interest, rather than other scales.

Globalised trends and narratives have affected how hydropower has developed in the MRB. In the 1960s and early 1970s, for example, dams were seen largely as symbols of post-colonial modernisation and anti-communist development (Sneddon, 2015). In the 1990s, following decades of Cold War conflict, dams returned to the region, sometimes being associated with its post-Cold War revitalisation and development (International Rivers Network, 1999). More recently, the political geographies have further shifted again with the geopolitical and economic strengthening of China, leading to new opportunities but also serious threats and challenges.

At the same time, dams are increasingly being defined by governments and lending bodies as 'renewable' or 'green' energy as a way to integrate dams into global climate mitigation schemes, such as the UN Clean Development Mechanism (e.g., Baird and Green, 2020). The narrative of renewable energy has had a huge positive political influence on the dam-building industry, demonstrating how political narratives can greatly affect development trajectories. Such 'sustainable hydropower' narratives are being increasingly offered (Middleton, 2022), not so much as a way of making hydropower development more just to both humans and nonhumans, but especially to justify continued and increasing flows of capital and electricity.

Conclusions

Political geographies help to ensure that hydropower development is not inappropriately depoliticised, but they are never easy to fully understand, as they relate to a range of values, priorities, interests, and scales. In this chapter, I have touched on some of the most important ways that different political geographies are creating complex hydropolitics assemblages, ones with particular anomalies and contradictions embedded within them.

There is potential for making development decision-making processes more inclusive, fair, and sensitive to poor and nonhuman actors, but much would need to change for such goals to truly manifest, and to support less influential people and species without voices. The countries that share the MRB would need to move away from politics-impeding 'sustainability' discourses, and more towards rights-based approaches. Time, however, is running out for the Mekong River and those who depend on it for their livelihoods. This does not mean that all is lost, or that nothing can be done. While the prospects for ideal outcomes may seem remote, substantial improvements are still possible, but the first step to finding appropriate solutions to these daunting challenges—even if they are imperfect—is recognising the nature of present and future problems, and their links to an array of different types of hydropolitics.

Finally, applying political geography as a conceptual tool can help us make links between politics, power, scale, and decision making so as to understand the circumstances in expansive and localised ways, ones that are attentive to a range of issues, and are appropriately nuanced. While it has not been possible to address all the relevant factors, including the important role that national politics and neopatrimonial political relations play in creating particular geographies, whether just or not, hopefully this chapter has made it clear that Mekong political geographies are important, even if they are also complex and constantly shifting in relation to the various types of flows that it has described.

References

Agnew, J. 1997. *Political geography: a reader*. London: Arnold.

Anderson, J. 2021 (revised). Assemblages. Oxford Bibliographies, https://10.1093/obo/9780199874002-0114.

Baird, I.G. 2016. Non-government organizations, villagers, political culture and the Lower Sesan 2 dam in northeastern Cambodia. *Critical Asian Studies* 48(2): 257–277, https://doi.org/10.1080/14672715.2016.1157958.

Baird, I.G. and Green, W.N. 2020. The Clean Development Mechanism and large dam development: contradictions associated with climate financing in Cambodia. *Climatic Change* 161: 365–383, https://doi.org/10.1007/s10584-019-02621-4.

Baird, I.G. and Hogan, Z.S. 2023. Hydropower dam development in the Mekong River Basin and fish biodiversity: a review. *Water* 15(7): 1352, https://doi.org/10.3390/w15071352.

Baird, I.G. and Manorom, K. 2019. Migrating fish and mobile knowledge: situated fishers' knowledge and social networks in the Lower Mekong River Basin in Thailand, Laos and Cambodia. *Mobilities* 14(6): 762–777, https://doi.org/10.1080/17450101.2019.1635343.

Baird, I.G. and Quastel, N. 2015. Re-scaling and reordering nature-society relations: the Nam Theun 2 hydropower dam and Laos-Thailand electricity networks. *Annals of the Association of the American Geographers* 105(6): 1221–1239, https://doi.org/10.1080/00045608.2015.1064511.

Baird, I.G., Flaherty, M.S. and Phylavanh, B. 2003. Rhythms of the river: lunar phases and migrations of small carps (Cyprinidae) in the Mekong River. *Natural History Bulletin of the Siam Society* 51(1): 5–36, https://thesiamsociety.org/wp-content/uploads/2020/04/NHBSS_051_1g_Baird_RhythmsOfThe-RiverL.pdf (accessed 16 July 2024).

Baird, I.G. and Flaherty, M.S. 2005. Mekong River fish conservation zones in southern Laos: assessing effectiveness using local ecological knowledge. *Environmental Management* 36(3): 439–454, https://doi.org/10.1007/s00267-005-3093-7.

Baird, I.G., Manorom, K., Phenow, A. and Gaja-Svasti, S. 2020. Opening the gates of the Pak Mun Dam: fish migrations, domestic water supply, irrigation projects and politics. *Water Alternatives* 13(1): 141–159, https://www.water-alternatives.org/index.php/alldoc/articles/vol13/v13issue1/568-a13-1-7/file (accessed 12 July 2024).

Bakker, K. 1999. The politics of hydropower: developing the Mekong. *Political Geography* 18(2): 209–232, https://doi.org/10.1016/S0962-6298(98)00085-7.

Barney, K. and Souksakoun, K. 2021. Credit crunch: Chinese infrastructure lending and Lao sovereign debt. *Asia & the Pacific Policy Studies* 8(1): 94–113, https://onlinelibrary.wiley.com/doi/10.1002/app5.318.

Bush, S.R. and Hirsch, P. 2005. Framing fishery decline. *Aquatic Resources, Culture and Development* 1(2): 79–90, https://doi.org/10.1079/ARC20048.

Chuenchum, P., Xu, M. and Tang, W. 2023. Assessment of reservoir trapping efficiency and hydropower production under future projections of sedimentation in Lancang–Mekong River Basin. *Renewable and Sustainable Energy Reviews* 184(2023): 113510, https://doi.org/10.1016/j.rser.2023.113510.

Dore, J. and Lebel, L. 2010. Deliberation and scale in Mekong Region water governance. *Environmental Management*, 46(1): 60–80, https://doi.org/10.1007/s00267-010-9527-x.

Eyler, B. and Weatherby, C. 2020. New evidence: how China turned off the tap on the Mekong River. Washington DC: Stimson Center, https://www.stimson.org/2020/new-evidence-how-china-turned-off-the-mekong-tap/ (accessed 2 June 2024).

Floch, P., Promphakping, B. and Blake, D.J.H. 2009. The 'greening of Isaan': politics, ideology and irrigation development in the northeast of Thailand. In Molle, F., Foran, T. and Käkönen, M. (eds.), *Contested waterscapes in the Mekong Region: hydropower, livelihoods and governance*, 253–282. London: Earthscan.

Foran, T. and Manorom, K. 2009. Pak Mun dam: Perpetually contested? In Molle, F., Foran, T. and Käkönen, M. (eds.), *Contested waterscapes in the Mekong Region: hydropower, livelihoods and governance*, 55–80. London: Earthscan.

Friend, R., Robert, A. and Keskinen, M. 2009. Songs of the doomed: the continuing neglect of capture fisheries in hydropower development in the Mekong. In Molle, F., Foran, T. and Käkönen, M. (eds.), *Contested waterscapes in the Mekong Region: hydropower, livelihoods and governance*, 307–333. London: Earthscan.

Fu, K.D., He, D.M. and Lu, X.X. 2008. Sedimentation in the Manwan reservoir in the Upper Mekong and its downstream impacts. *Quaternary International* 186(1): 91–99, https://doi.org/10.1016/j.quaint.2007.09.041.

Geheb, K. and Suhardiman, D. 2019. The political ecology of hydropower in the Mekong River Basin. *Current Opinion in Environmental Sustainability* 37: 8–13, https://doi.org/10.1016/j.cosust.2019.02.001.

Glassman, J. 2002. From Seattle (and Ubon) to Bangkok: the scales of resistance to corporate globalization. *Environment and Planning D: Society and Space* 20(5): 513–533, https://doi.org/10.1068/d334.

Global Witness 2016. Hostile takeover: the corporate empire of Cambodia's ruling family. London: Global Witness, https://gw.hacdn.io/media/documents/Hostile_Takeover_Cambodia_report_July_16.pdf (accessed 24 June 2026).

Goldman, M. 2005. *Imperial nature: the World Bank and struggles for Justice in the age of globalization*. New Haven (CT): Yale University Press.

Green, W.N. and Baird, I.G. 2016. Capitalizing on compensation: Hydropower resettlement and the commodification and decommodification of nature–society relations in Southern Laos. *Annals of the Association of American Geographers* 106(4): 853–873, https://doi.org/10.1080/24694452.2016.1146570.

Green, W.N. and Baird, I.G. 2020. The contentious politics of hydropower dam impact assessments in the Mekong River basin. *Political Geography* 83(2020): 102272, https://doi.org/10.1016/j.polgeo.2020.102272.

Grünwald, R., Feng, Y. and Wang, W. 2021. Politicization of science in the Lancang–Mekong Basin: the Eyes on Earth study. *International Journal of Water Resources Development* 39(2): 184–210, https://doi.org/10.1080/07900627.2021.1990025.

Hecht, J.S., Lacombe, G., Arias, M.E., Thanh Duc Dang and Piman, T. 2019. Hydropower dams of the Mekong River basin: A review of their hydrological impacts. *Journal of Hydrology* 568(2019): 285–300, https://doi.org/10.1016/j.jhydrol.2018.10.045.

Hennig, T., Wang, W., Magee, D. and He, D. 2016. Yunnan's fast-paced large hydropower development: a powershed-based approach to critically assessing generation and consumption paradigms. *Water* 8(10): 476, https://doi.org/10.3390/w8100476.

Hensengerth, O. 2015. Global norms in domestic politics: environmental norm contestation in Cambodia's hydropower sector. *The Pacific Review* 28(4): 505–528, https://doi.org/10.1080/09512748.2015.1012107.

Hirsch, P. 2007. Civil society and interdependencies: towards a regional political ecology of Mekong development. In Connel, J. and Waddell, E. (eds.), *Environment, development and change in rural Asia-Pacific: between local and global*, 226–246. New York (NY): Routledge.

Hirsch, P. 2010. The changing political dynamics of dam building on the Mekong. *Water Alternatives* 3(2010): 312–323, https://www.water-alternatives.org/index.php/volume3/v3issue2/95-a3-2-18/file (accessed 16 July 2024).

Hirsch, P. 2011. China and the cascading geopolitics of Lower Mekong Dams. *Asia-Pacific Journal* 9(20): 1–5, https://apjjf.org/2011/9/20/Philip-Hirsch/3529/article (accessed 14 October 2023).

Hirsch, P. 2016. The shifting regional geopolitics of Mekong Dams. *Political Geography*, 51(2016): 63–74, https://doi.org/10.1016/j.polgeo.2015.12.004.

Hirsch, P. and Wyatt, A. 2004. Negotiating local livelihoods: scales of conflict in the Se San River basin. *Asia Pacific Viewpoint* 45(1): 51–68, https://doi.org/10.1111/j.1467-8376.2004.00227.x.

IRN (International Rivers Network). 1999. Power struggle: the impacts of hydro-development in Laos. Berkeley (CA): International Rivers Network, https://riverresourcehub.org/resources/power-struggle-the-impacts-of-hydro-development-in-laos-4058/ (accessed 29 March 2026).

International Rivers. 2008. Power surge: the impacts of rapid dam development in Laos. Berkeley (CA): International Rivers, https://riverresourcehub.org/wp-content/uploads/files/attached-files/intl_rivers_power_surge.pdf (accessed 29 March 2026).

Jessop, B., Brenner, N. and Jones, M. 2008. Theorizing sociospatial relations. *Environment and Planning D: Society and Space* 26(3): 389–401, https://doi.org/10.1068/d9107.

Käkönen, M. and Hirsch, P. 2009. The anti-politics of Mekong knowledge production. In Molle, F., Foran, T. and Käkönen, M. (eds.), *Contested Waterscapes in the Mekong Region: hydropower, livelihoods and governance*, 333–365. London: Earthscan.

Käkönen, M. and Nygren, A. 2023. Resurgent dams: Shifting power formations, persistent harms, and obscured responsibilities, *Globalizations* 20(6): 866–886, https://doi.org/10.1080/14747731.2022.2098668.

Käkönen, M. and Thuon, T. 2019. Overlapping zones of exclusion: carbon markets, corporate hydropower enclaves and timber extraction in Cambodia. *Journal of Peasant Studies* 46(6): 1192–1218, https://doi.org/10.1080/03066150.2018.1474875.

Kottelat, M., Baird, I.G., Kullander, S.O., Ng, H.H., Parenti, L.R., Rainboth, W.J. and Vidthayanon, C. 2012. The status and distribution of freshwater fishes of Indo-Burma. In Allen, D.J., Smith, K.G. and Darwell, W.R.T. (eds.), *The status and distribution of freshwater biodiversity in Indo-Burma*, 36–65. Cambridge and Gland: International Union for the Conservation of Nature.

Kripakorn, D. 2007. Social movement and the natural resources management: a study of the anti Nam chon dam movement (1982-1988). *Silpakorn University International Journal* 7(2007): 112–142, https://www.thaiscience.info/Journals/Article/SUIJ/10499120.pdf (accessed 17 July 2024).

Kurtz, H.E. 2003. Scale frames and counter-scale frames: constructing the problem of environmental injustice. *Political Geography* 22(8): 887–916, https://doi.org/10.1016/j.polgeo.2003.09.001.

Lamb, V. 2014. Making governance 'good': the production of scale in the environmental impact assessment and governance of the Salween River. *Conservation and Society* 12(4): 386, https://doi.org/10.4103/0972-4923.155582.

Lebel, L. 2006. The politics of scale in environmental assessments. In Reid, W.V., Berkes, F., Wilbanks, T. and Capistrano, D. (eds.), *Bridging scales and knowledge systems: concepts and applications in ecosystem assessment*, 37–58. Washington (DC): Island Press.

Leitner, H., Sheppard, E. and Sziarto, K.M. 2008. The spatialities of contentious politics. *Transactions of the Institute of British Geographers* 33(2): 157–172, https://doi.org/10.1111/j.1475-5661.2008.00293.x.

Luukkanen, J., Tuominen V. and Vehmas, J. 2012. Scales and fields of electricity production: sustainability discourses of electricity production in Cambodia and Laos. *Forum for Development Studies* 39(2): 209–230, https://doi.org/10.1080/08039410.2012.666671.

Marks, D. and Zhang, J. 2019. Circuits of power: Environmental injustice from Bangkok's shopping malls to Laos' hydropower dams. *Asia Pacific Viewpoint* 60(3): 296–309, https://doi.org/10.1111/apv.12242.

McCully, P. 2001. *Silenced rivers: the ecology and politics of large dams* (Enlarged & Updated). London: Zed Books.

Magee, D. 2006. Powershed politics: hydropower and interprovincial relations under Great Western Development. *The China Quarterly* 185: 23–41, https://doi.org/10.1017/S0305741006000038.

Mahanty, S., Milne, S., Barney, K., Dressler, W., Hirsch, P. and To, P.X. 2023. Rupture: towards an emplaced, critical, and experiential view of nature-society crisis. *Dialogues in Human Geography* 13(2): 177–196, https://doi.org/10.1177/20438206221138057.

Matthews, N. 2013. Drivers and enablers of hydropower development in the Lao PDR: the political ecology of Mekong riparians, investors and the environment. PhD Thesis. London: Department of Geography, King's College London, University of London, https://kclpure.kcl.ac.uk/portal/files/103927104/2013_Matthews_Nathanial_1018111_ethesis.pdf (accessed 5 February 2023).

Middleton, C. 2022. The political ecology of large hydropower dams in the Mekong Basin: a comprehensive review. *Water Alternatives* 15(2): 251–289, https://www.water-alternatives.org/index.php/alldoc/articles/vol15/v15issue2/668-a15-2-10/file (accessed 6 May 2024).

Middleton, C. and Allouche, J. 2016. Watershed or powershed? A critical hydropolitics of the 'Lancang-Mekong Cooperation Framework'. *The International Spectator* 51(3): 100–117, https://doi.org/10.1080/03932729.2016.1209385.

Middleton, C. and Pritchard, A. 2016. Arenas of water justice on transboundary rivers: a case study of the Xayaburi Dam, Laos. In Blake, D.J.H. and Robins, L. (eds.), *Water governance dynamics in the Mekong Region*, 55–90. Petaling Jaya: Strategic Information & Research Development Centre.

Middleton, C., Garcia, J. and Foran, T. 2009. Old and new hydropower players in the Mekong Region: agendas and strategies. In Molle, F., Foran, T. and Käkönen, M. (eds.), *Contested waterscapes in the Mekong Region: hydropower, livelihoods and governance*, 23–54. London: Earthscan.

Middleton, C., Grundy-Warr, C. and Yong, M.L. 2013. Neoliberalizing hydropower in the Mekong Basin: the political economy of partial enclosure. *Social Science Journal* 43(2): 299–334, https://www.academia.edu/11708971/Neoliberalizing_Hydropower_in_the_Mekong_Basin_The_Political_Economy_of_Partial_Enclosure (accessed 4 December 2024).

Middleton, C., Allouche, J. Gyawali, D. and Allen, S. 2015a. The rise and implications of the water-energy-food nexus in Southeast Asia through an environmental justice lens. *Water Alternatives* 8(1): 627–654, https://www.water-alternatives.org/index.php/all-abs/269-a8-1-2/file (accessed 27 July 2024).

Middleton, C., Matthews, N. and Mirumachi, N. 2015b. Whose risky business? Public-Private Partnerships (PPP), Build-Operate-Transfer (BOT) and large hydropower dams in the Mekong Region. In Matthews, N. and Geheb, K. (eds.), *Hydropower development in the Mekong Region: political, socio-economic and environmental perspectives*, 127–152. London: Earthscan.

Missingham, B.D. 2003. *The Assemblies of the Poor in Thailand: from local struggles to national protest movement.* Chiang Mai: Silkworm Books.

Orr, S., Pittock, J., Chapagain, A. and Dumaresq, D. 2012. Dams on the Mekong River: lost fish protein and the implications for land and water resources. *Global Environmental Change* 22(4): 925–932, https:// doi.org/10.1016/j.gloenvcha.2012.06.002.

Porter, I.C. and Shivakumar, J. 2011. *Doing a dam better: the Lao People's Democratic Republic and the story of the Nam Theun 2*. Washington (DC): The World Bank.

Rainboth, W.J. 2012. Fishes of the Greater Mekong ecosystem with species list and photographic atlas. Miscellaneous Publications of the University of Michigan Museum of Zoology. Ann Arbor (MI), https://deepblue.lib.umich.edu/bitstream/handle/2027.42/122721/MP%20201sm.pdf?sequence=2&isAllowed=y (accessed 16 July 2024).

Scudder, T. 2005. *The future of large dams: dealing with social, environmental, institutional and political costs*. London: Earthscan.

Shoemaker, B. and Robichaud, W. (eds.). 2018. *Dead in the water: global lessons from the World Bank's model hydropower project in Laos*. Madison (WI): University of Wisconsin Press.

Silvano, R.A.M., Baird, I.G., Begossi, A., Hallwass, G., Huntington, H.P., Lopes, P.F.M., Parlee, B. and Berkes, F. 2023. Fishers' multidimensional knowledge advances in fisheries and aquatic science. *Trends in Ecology and Evolution* 38(1): 8–12, https://doi.org/10.1016/j.tree.2022.10.002.

Sneddon, C. 2003. Reconfiguring scale and power: the Khong-Chi-Mun Project in northeast Thailand. *Environmental Planning* 35(2003): 2229–2250, https://10.1068/a35299.

Sneddon, C. 2015. *Concrete revolution: large dams, cold war geopolitics, and the US Bureau of Reclamation*. Chicago (IL): University of Chicago Press.

Soukhaphon, A., Baird, I.G. and Hogan, Z. 2021. Hydropower dams and impacts in the Mekong River Basin: a review. *Water* 13(3): 265, https://doi.org/10.3390/w13030265.

South China Morning Post 2011. Controversial Chinese projects in Cambodia bow to public pressure. *South China Morning Post*, 3 September 2011, https://www.scmp.com/article/977985/controversial-chinese-projects-cambodia-bow-public-pressure (accessed 12 September 2024).

Strangio, S. 2023. Cambodian PM Affirms Ban on Mekong Hydropower Projects. *The Diplomat* 1 December 2023, https://thediplomat.com/2023/12/cambodian-pm-affirms-ban-on-mekong-hydropower-projects/ (accessed 2 July 2024).

Sultana, F. 2018. Water justice: why it matters and how to achieve it. *Water International* 43(4): 483–493, https://doi.org/10.1080/02508060.2018.1458272.

Suhardiman, D., Giordano, M. and Molle, F. 2011. Scalar disconnect: the logic of transboundary water governance in the Mekong. *Society & Natural Resources* 25(6): 572–586, https://doi.org/10.1080/08941920.2011.604398.

Urban, F. 2013. An analysis of China's investment in the hydropower sector in the Greater Mekong Sub-Region. *Environment, Development and Sustainability* 15(2): 1–24, https://doi.org/10.1007/s10668-012-9415-z.

WCD (World Commission on Dams). 2000. *Dams and development: a new framework for decision-making*. London: Earthscan Publications.

Whitington, J. 2018. *anthropogenic rivers: the production of uncertainty in Lao hydropower*. Ithaca (NY): Cornell University Press.

Wyatt, A.B. 2004. Infrastructure development and BOOT in Laos and Vietnam: a case study of collective action and risk in transitional developing economies. PhD Thesis. Sydney: The University of Sydney.

Wyatt, A.B. and Baird, I.G. 2007. Transboundary impact assessment in the Sesan River basin: the case of the Yali Falls dam. *International Journal of Water Resources Development* 23(3): 427–442, https://doi.org/10.1080/07900620701400443.

Xinhua. 2019. Laos seeks to become regional hub for electricity transmission by 2025. *Xinhua* 15 February 2019, http://www.xinhuanet.com/english/asiapacific/2019-02/14/c_137821006.htm (accessed 21 August 2023).

Acknowledgements

Each chapter in this volume has been peer reviewed by at least three reviewers, one of whom was an editor. The other two (or more) were anonymous.

The editors and chapter authors would like to acknowledge, with sincere thanks, the time, support, and efforts provided by the reviewers listed below. Final responsibility for the chapters remains with their authors.

Keith Barney, Associate Professor, Resources, Environment and Development Department, Crawford School of Public Policy, Australian National University, Canberra, Australia.

Sebastian Biba, Senior Researcher, Professorship for Political Science, Institute for Political Science, Faculty 03 Social Science, Goethe University, Frankfurt, Germany.

Mark Bo is an independent researcher, who has lived and worked in Southeast Asia for two decades, working on corporate accountability and the impacts of large-scale development projects.

Saturnino ('Jun') M. Borras Jr., Professor of Agrarian Studies, International Institute of Social Studies, Erasmus University Rotterdam, The Netherlands.

David J. Devlaeminck, Associate Professor, School of Law, Chongqing University, People's Republic of China

John Dore, Lead Specialist, Climate Resilience & Water Security, Australia's Department of Foreign Affairs and Trade, Bangkok, Thailand.

Phil Hirsch, Professor Emeritus of Human Geography, University of Sydney; and Research Affiliate, Regional Center for Social Science and Sustainable Development, Chiang Mai University, Thailand.

Mira Käkönen, ARC DECRA Fellow, Resources, Environment & Development Group, Crawford School of Public Policy, Australian National University, Canberra, Australia.

Darrin Magee, Director, Institute of Energy Studies, Western Washington University, Bellingham, Washington, USA.

Melissa Marschke, Professor, School of International Development and Global Studies, University of Ottawa, Ontario, Canada.

Miles Kenney-Lazar, Senior Lecturer in Human Geography, School of Geography, Earth and Atmospheric Sciences, University of Melbourne, Australia.

Naho Mirumachi, Professor in Environmental Politics, Department of Geography, King's College, London, UK.

Wigke Sukmana Putri, PhD Researcher, The Institute of Cultural Anthropology and Development Sociology and the Royal Netherlands Institute of Southeast Asian and Caribbean Studies, Leiden University, The Netherlands.

Jonathan Rigg, Professor, Chair of Geographical Science, University of Bristol, UK.

Kanya Souksakoun, Deputy Director, Environmental Management Department, Faculty of Environmental Sciences, National University of Laos, Vientiane, Laos.

Olivier Tappe, Senior Researcher, Department of Anthropology, University of Heidelberg, Germany.

Chu Thai Hoanh, Emeritus Scientist, International Water Management Institute, Hanoi, Vietnam.

Widya Naseva Tuslian, PhD candidate, Leiden Law School, Leiden University, The Netherlands.

John Ward, Principal Researcher, Mekong Region Futures Institute, Vientiane, Laos.

Kevin Woods, Fellow, East-West Center, Honolulu, Hawai'i, USA and Research Associate, Faculty of Political Science, Chulalongkorn University, Bangkok, Thailand.

About the authors

Ian G. Baird (ibaird@wisc.edu) is a Professor at the Department of Geography, University of Wisconsin-Madison in the USA. His work focuses on, amongst others, political ecology and human–environment relations, mainly in Laos, Thailand, and northeastern Cambodia. He has published extensively on hydropower, fisheries, indigeneity, and other topics.

Saw John Bright (hserdohtoo@gmail.com) is the Water Governance Programme Coordinator at the Karen Environmental and Social Action Network, Chiang Mai, Thailand. He is currently doing his PhD at Wageningen University, The Netherlands.

Tony Dinh (tdinh038@uottawa.ca) is a PhD student in the School of International Development and Global Studies at the University of Ottawa. His research interest centres around the environmental and socio-economic implications of sand extraction in the Red River Delta, Vietnam.

Genevieve Donnellon-May (genevieve.donnellonmay@gmail.com) is a Non-resident Fellow at the Pacific Forum (Honolulu, USA) and Associated Fellow at the Institute for Security and Development Policy (Stockholm, Sweden).

Kim Geheb (k.geheb@landscapealliance.org) is a geographer and Senior Researcher at the Landscape Alliance. His research interests focus on political ecologies, and the management and governance of natural resources systems. He is based in Nairobi, Kenya.

Carl Grundy-Warr (geocerg@nus.edu.sg) is a Senior Lecturer in the Department of Geography, Faculty of Arts and Social Sciences, National University of Singapore. He has extensive experience in the Mekong and mainland SEA, where his research focuses on political geographies and political ecologies of transboundary natural resources.

Kanokwan Manorom (kanokwan.m@ubu.ac.th) is Professor of Rural Sociology at Ubon Ratchathani University (Thailand). She researches political economy, natural resource governance, livelihoods, gender and social impact assessment in Northeast Thailand and the Mekong region, with over 25 years' experience in international research collaboration.

Casper Palmano (casper.palmano@gmail.com) is a Programme Advisor to the Karen Environmental and Social Action Network. He is a land tenure specialist, indigenous activist, and researcher. He is based in Hpa An, Myanmar.

Jean-François Rousseau (jf.rousseau@uottawa.ca) is an associate professor in the School of International Development and Global Studies at the University of Ottawa. His research focuses on the political projects that shape water and sand resource reallocations, and on the resulting livelihood implications. He conducts research in Southwest China and mainland Southeast Asia.

Thong Anh Tran (thong.tran@anu.edu.au) is a human ecologist. He is an Honorary Lecturer at Fenner School of Environment and Society, The Australian National University; and Researcher with the Mekong Institute, Can Tho University, Can Tho City, Vietnam. His research focuses on climate-development dynamics in the Mekong Region, with a particular focus on water governance and agrarian transition, amongst others.

Le Anh Tuan (latuan@ctu.edu.vn) is a Senior Lecturer at the College of Environment and Natural Resources, Can Tho University in Vietnam, and a Scientific Advisor to the Mekong Institute, Can Tho University. Vietnam. His research focuses on environmental hydrology, climate change, and water resources development

Mak Sithirith (sithirith.mak@cdri.org.kh) is a Scientist and Director of the Centre of Natural Resources and Environment and the Cambodian Development Resource Institute in Phnom Penh, Cambodia. He is a geographer and political ecologist, with a long specialisation in the fisheries of the Tonle Sap Great Lake.

Danzig Sopera (danzig.sopera@gmail.com) is the *nom de plume* of a researcher who prefers to remain anonymous. He is a geographer with a PhD from a European university. His extensive work in the Mekong Region has focused on the political ecologies of water, hydropower, and corruption.

Diana Suhardiman (suhardiman@kitlv.nl) is Director at the Royal Netherlands Institute of Southeast Asian and Caribbean Studies and Professor at Leiden University, The Netherlands. Her work has focused on scale, politics, and natural resources in Laos and Myanmar.

Ming Li Yong (mlyong@suss.edu.sg) is a Lecturer at the Singapore University of Social Sciences. Her research interests are in political ecology, transboundary water governance, natural resource management, and agrarian change in the Mekong region. She also leads collaborative projects foregrounding community perspectives on environmental change and vulnerability in Southeast Asia.

Hongzhou Zhang (ishzzhang@ntu.edu.sg) is an Assistant Professor at the S. Rajaratnam School of International Studies, Nanyang Technological University, Singapore. He received his PhD in Public Policy from the Lee Kuan Yew School of Public Policy, National University of Singapore. His main research interests include regional and global resources, conflicts and governance, game theory, discourse analysis, and the governance of emerging technologies.

Index

FSC
www.fsc.org
MIX
Papier | Fördert
gute Waldnutzung
FSC® C083411

Kolibri 360 GmbH
Ferdinand-Jühlke-Straße 7
99095 Erfurt, Deutschland
produktsicherheit@kolibri360.de